Ivo Boniolo
Sergio Matteo Savaresi

Estimate of the

GW01464260

Ivo Boniolo
Sergio Matteo Savaresi

Estimate of the lean angle of motorcycles

Design and analysis of systems for measuring and estimating the attitude parameters of motorcycles

VDM Verlag Dr. Müller

Impressum/Imprint (nur für Deutschland/ only for Germany)

Bibliografische Information der Deutschen Nationalbibliothek: Die Deutsche Nationalbibliothek verzeichnet diese Publikation in der Deutschen Nationalbibliografie; detaillierte bibliografische Daten sind im Internet über http://dnb.d-nb.de abrufbar.
Alle in diesem Buch genannten Marken und Produktnamen unterliegen warenzeichen-, marken- oder patentrechtlichem Schutz bzw. sind Warenzeichen oder eingetragene Warenzeichen der jeweiligen Inhaber. Die Wiedergabe von Marken, Produktnamen, Gebrauchsnamen, Handelsnamen, Warenbezeichnungen u.s.w. in diesem Werk berechtigt auch ohne besondere Kennzeichnung nicht zu der Annahme, dass solche Namen im Sinne der Warenzeichen- und Markenschutzgesetzgebung als frei zu betrachten wären und daher von jedermann benutzt werden dürften.

Coverbild: www.ingimage.com

Verlag: VDM Verlag Dr. Müller Aktiengesellschaft & Co. KG
Dudweiler Landstr. 99, 66123 Saarbrücken, Deutschland
Telefon +49 681 9100-698, Telefax +49 681 9100-988
Email: info@vdm-verlag.de
Zugl.: Milan, Politecnico di Milano, 2010

Herstellung in Deutschland:
Schaltungsdienst Lange o.H.G., Berlin
Books on Demand GmbH, Norderstedt
Reha GmbH, Saarbrücken
Amazon Distribution GmbH, Leipzig
ISBN: 978-3-639-26328-2

Imprint (only for USA, GB)

Bibliographic information published by the Deutsche Nationalbibliothek: The Deutsche Nationalbibliothek lists this publication in the Deutsche Nationalbibliografie; detailed bibliographic data are available in the Internet at http://dnb.d-nb.de.
Any brand names and product names mentioned in this book are subject to trademark, brand or patent protection and are trademarks or registered trademarks of their respective holders. The use of brand names, product names, common names, trade names, product descriptions etc. even without a particular marking in this works is in no way to be construed to mean that such names may be regarded as unrestricted in respect of trademark and brand protection legislation and could thus be used by anyone.

Cover image: www.ingimage.com

Publisher: VDM Verlag Dr. Müller Aktiengesellschaft & Co. KG
Dudweiler Landstr. 99, 66123 Saarbrücken, Germany
Phone +49 681 9100-698, Fax +49 681 9100-988
Email: info@vdm-publishing.com

Printed in the U.S.A.
Printed in the U.K. by (see last page)
ISBN: 978-3-639-26328-2

To Ilaria (I.B.)
To Cristina, Claudio and Stefano (S.M.S)

Contents

List of figures

List of tables

Chapter 1
Introduction

The necessity of measuring and estimate a parameter is important to satisfy the knowledge about the condition in which a system is operating. In particular, in the modern time, to close a feedback loop on a system it is necessary to have information about the state of the controlled plant. Talking about vehicle, both in automotive context and in nautical application, it is necessary to know the attitude of the means of transportation to correctly act on it.

This work is devoted to the estimation of attitude parameters in vehicles. This problem is not trivial: as in all engineering applications, it is necessary to determine the correct trade-off between the costs and the performances, and this trade-off also depends on the operating filed. Two main areas have to be distinguished:

- *Racing competitions*: many instruments are available on the market to measure with high precision most of the interesting parameters of a vehicle. These devices often require an high mounting reliability and cannot be adopted in a consumer application both because of the expensiveness and due to the inapplicability on a common vehicle. The application justify these costs, but it is obviously impossible to sustain them on the consumer market;

- *Consumer market*: in this case the target is to minimize the costs and the system complexity, thus, the foremost objective is the reduction of the number of sensors and the usage of very low cost devices. Therefore, the accuracy of the reconstructed parameters cannot be as good as in a racing application and it is has just to be sufficiently accurate for the specific problem. For example, the stability control problem on a consumer car is based on the signal provided by a low cost yaw gyroscope.

The measurement and the estimation of the attitude of the motorbike is important both in racing and consumer field, as a consequence, two different solutions are proposed: one more expensive that guarantees high precision (less than 1°) and one based on very low cost sensors that assures less accuracy (around 5°).

1.1 Aim of the book and book organization

The aim of this book and how is been organized is discussed herein. The estimation of the attitude of a vehicle is a well known problem in the scientific literature. As a consequence, Chapter 2 is devoted to the analysis of the main contributions that can be found in the literature and to the introduction to the topic that is treated in this book.

Regarding the lean angle estimation of a motorcycle, on the scientific literature just minor contributions can be found, while it is a very discussed problem in the patent literature. This fact underlines the importance of having a reliable estimation of the attitude of a motorcycle especially in the industrial applications. Electro-optical and inertial solutions will be studied to solve the problem both for racing and industrial applications;

In estimating the attitude of a vehicle, most of the times, it is necessary to relate to inertial sensors such as accelerometers and gyroscopes. The quantities that are provided by these sensors are kinematic, thus, it is important to recall some basic concepts:

- Rotational matrices;
- Euler angles;
- Kinematic relations.

All of these arguments are treated in Chapter 3, in which also the Kalman filtering approach and Neural Networks are described, because of their importance in the attitude estimation and calibration framework. These tools are also used in this book to study the problem of lean angle estimation.

Figure 1.1: Leaned motorcycle in race competition.

The problem that is tackled in the book is the roll angle estimation of a motorcycle (Figure 1.1), that is the foremost attitude parameter that influence the state of a two-wheeled vehicle. In this book, from Chapter 4 to Chapter 9 is devoted to this topic.

First of all in Chapter 4 the some preliminaries on the roll angle estimation are introduced. In particular, the content is focused on the definition of the lean angle of a two wheeled vehicle defining:

- the Euler roll angle: inclination of the symmetry plane of the vehicle with respect to the absolute vertical plane;
- the road roll angle: inclination of the symmetry plane of the vehicle with respect to the plane perpendicular to the road;
- the inertial roll angle: inclination of the plane containing the tires contact point and the Center Of Gravity (COG) of the system rider-motorcycle with respect to the absolute vertical plane.

The end of Chapter 4 is devoted to the description of the simulation environment and the experimental set-up that have been used to validate the proposed measurement and estimation systems.

The highest precision system to measure the inclinations of the motorcycle with respect to the road is the electro-optical system. The electro-optical sensors are the unique devices that are able to separate the road bank contribution from the inclination of the two-wheeled vehicle. The drawback of the sensors that are available on the market is that they are designed to be mounted on a car, as a consequence:

- Electro-optical sensors of-the-shelf are too much expensive for a motorcycle application;
- Available sensors are difficult to handle because of their dimension and cannot be easily mounted on a motorbike;
- The electro-optical measuring sensors that are used in cars to measure the distance between chassis and road have an unnecessarily high precision for a motorcycle application.

In this work the problem of designing and testing of a new electro-optical component for motorcycle application is studied. This component is fundamental especially in the racing context and to correctly reconstruct the benchmark signal to be used in the designing of estimation algorithms. In Chapter 5 the sensor in characterized and different problems are addressed:

- The interaction between the vehicle speed and the road roughness decreases the Signal to Noise ratio of the sensor, thus, the source of light of the telemeter needs to be carefully defined and the signal needs to be accurately filtered;
- The dynamic response of the sensor and the effect of the tires thickness is studied;
- The accuracy of the system strongly depends on the mounting position of the sensors, thus, the measurement precision is studied to define the best mounting position and to show that an high precision electro-optical sensor is not necessary to obtain a roll angle accuracy less than $1°$.

Once a system to measure in real time the attitude angles of the vehicle is realized, the estimation problem is taken into account. The objective is to use low-cost inertial sensors to recover the attitude angles of the motorcycle.

Just on the top of the racing competitions (e.g. MotoGP and Superbike teams), some motorcycles are equipped with inertial sensors and just few of them use inertial signals to close a feedback control loop. Most of them found the control system on the signals provided by a GPS system that are also adopted to reconstruct the lean angle of the vehicle: this position signals based algorithm is described in Chapter 6. The algorithm is based on:

- the definition of the lean angle of the vehicle as the quantity that is necessary to maintain an equilibrium between the gravitational acceleration and the centrifugal acceleration;
- the estimation of the curvature radius of the path through position signals.

The main drawback of this estimation method is that the accuracy strongly depends on the availability and precision of the GPS signals and it is well known that the position signals are characterized by a time lag (around 0.5 s) and they suffer of the close sky condition (absence of GPS signals).

In racing competitions it is desired to have an independent unit which mounting position is not known a priori, on the contrary, in consumer applications the position and the orientation of the unit is fixed by manufacturers. Thus, in general, four main topics have to be considered to solve the lean angle estimation problem:

- Definition of the signals models;
- Alignment of the sensors in the body reference frame;
- Offset estimation;
- Attitude estimation.

In Chapter 7, an high level description of the adopted approach to solve the problem is given. In particular, after the presentation of the models that are used in the design of the attitude estimation algorithms, it is highlighted that the problems of signals alignment, offset estimation and attitude estimation are solved independently, so that the overall complexity is split and the three objective are separately treated. An optimal algorithm to align the acquired signals in the body reference frame is proposed: this is important because of the fact that the developed models are referred to quantity acquired in the body coordinate system of the vehicle. The bias of the signals can be estimate with a grey-box method based on the implementation of a double buffer algorithm.

The problem of attitude estimation is taken into account in Chapter 8 and Chapter 9.

Chapter 8 is devoted to the presentation of the frequency separation based algorithms in which the roll angle of the vehicle is estimated as the sum of two component:

- High frequency component obtained integrating the high frequency contribution of the roll gyroscope,
- Low frequency component obtained by the application of static expression.

The design of the low frequency estimation algorithms is the main argument of the Chapter. The estimation performances are firstly analyzed in a Neural Network framework to show that through the estimation of the yaw rate of the vehicle it is possible to reach the best performance. Starting from this conclusion, different low frequency estimation algorithms are compared analyzing the estimation performance both in a simulated and experimental environment. This algorithms are based on the expression of the inertial roll angle of a two wheeled vehicle and they results easy to be implement requiring just a subset of the signals provided by a IMU that contains three accelerations and three angular rates.

Chapter 9 is devoted to the application of Kalman filtering approaches for estimating the attitude of the motorcycle. In a motorcycle application it is not necessary to obtain very high estimation accuracy, but it is important to have an high bandwidth of the estimator, thus a direct formulation of the Kalman Filter is proposed to explicitly estimate the attitude of the vehicle. The description of the process is discussed and the attitude of the vehicle is estimated with an implementation of the Extended Kalman Filter and also implementing the Unscented Kalman Filter. In a Kalman filtering framework it is possible to estimate both the pitch angle and the roll angle of the vehicle. Because of the estimation of the pitch dynamic of the vehicle the estimation accuracy is better than the precision of the frequency separation based algorithms. Moreover, it is shown that in the motorcycle context:

- the EKF and the UKF are completely equivalent both on the accuracy point of view and on the robustness point of view;
- the estimation of the signals errors in a Kalman filtering framework instead of using the grey-box offset observer does not improve the estimation accuracy.

Finally, in Chapter 10 some conclusions and future works are summarized.

1.2 Main contributions

The main contributions of the book can be summarized as follows.

- A novel sensor has been designed to measure the attitude angles of a motorbike reliably and with high accuracy (Chapter 5);
- A new approach to solve the problem of attitude estimation via inertial sensors has been introduced (Chapter 7), in particular:
    o The problem of alignment of the inertial signals in the body coordinate system is optimally solved;
    o A novel offset grey-box offset observer is introduced;
- A set of new frequency based algorithms is introduced to estimate the lean angle of the vehicle with a few numbers of sensors and an ESR (Error to Signal Ratio) of about 5% (Chapter 8), in particular:
    o The importance of the yaw rate estimation is underlined;
    o The effect of the pitch dynamic is studied;
    o The decrease of performance due to banked road is presented.
- The attitude of the motorcycle is studied in a Kalman filtering framework and with a complete IMU it is possible to reach an ESR of about 2.5% (Chapter 9):
    o A new description of the process is proposed to directly estimate the attitude of the motorbike;
    o The EKF and the UKF are compared to show that in a motorcycle application they are equivalent;
    o An augmented description of the state of the process with the signals errors is analyzed to show that the increase of the complexity is not justified by an increase of the estimation accuracy.
- The performance improvement due to the KF approach is studied showing that it is due just to the estimation of the pitch dynamic of the vehicle and that the presence of a banked road causes an additive error that cannot be compensated by inertial measurements.

1.3 Disclaimer

Some of the presented algorithms are protected by Italian patent and International patent

Chapter 2
State of the art and problem definition

This Chapter is devoted to the presentation of the state of the art of the estimation topic treated in this book.

Nowadays, four-wheeled vehicles are equipped with many different active control systems which enhance driver and passengers safety, some of which - such as the ABS (Anti-lock Braking System) - have recently become a standard on all cars (see *e.g.*, [1], [2]).

In the field of two-wheeled vehicles, instead, electronic systems for the control of vehicle dynamics are still in their infancy: only a few motorbikes are equipped with ABS systems; traction control, drive-by-wire, and semi-active suspensions are mostly confined to advanced R&D prototypes and racing competitions; Electronic Stability Control (ESC) systems are still far from an industrial application and only preliminary studies have been done on this topic (see e.g. the EU-sponsored REGINS project "SAFEBIKE" - http://safebike.jku.at).

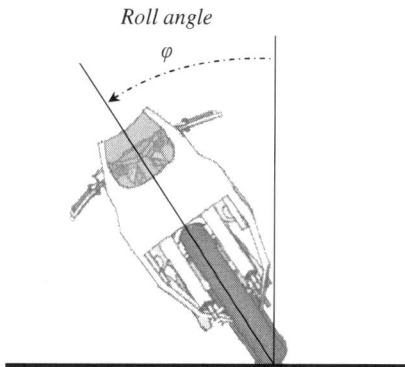

Figure 2.1: Roll angle of a two-wheeled vehicle.

More specifically, the ABS systems today available on the market are expected to work at their best only when panic brakes occurs at in-plane conditions (zero roll angle). However, it is well known that a two-wheeled vehicle is characterized by large values of *roll angle* (see Figure 2.1), which can reach the astonishing value of 50°-55° using high-performance racing tyres. This angle is the inclination of the vehicle with respect to the vertical direction and it represents the amount of inclination that the bike needs in order to ensure the force balance on the curve. These large roll angles obviously play a major role in the overall vehicle dynamics, and make the motorcycle dynamics very different (and much more complicated to be modelled and controlled) from the car dynamics.

Hence, to move a step further in active control systems for two-wheeled vehicle dynamics, the enabling technology comes from a system capable to estimate the roll angle in a reliable way and in real-time. Moreover, to suit industrial cost constraints, such system should rely on a low-cost sensor configuration.

Besides its usefulness in control system design, a reliable on-line measure of the roll angle might be useful in the racing context to assess the tyre performance with respect to this fundamental variable. In fact (see *e.g.*, [3], [4], [5], [6], [7]) the roll angle has a major impact in determining the lateral tyre-road contact forces.

In the open scientific literature very little has been published on this topic so far. In [8], an interesting approach for estimating the whole vehicle trajectory is proposed; it employs a vision system made by cameras complemented with MEMS accelerometers.

Something more can be found in the recent patent literature. For example, in [9], [10], [11] some approaches are described, whose common purpose is to devise robust estimation methods with low cost (and small size) equipment. Specifically, [11] focuses on roll-over detection mainly tailored to four-wheeled vehicles while [9], [10] focus on accelerometer-based estimation algorithms.

In the literature about attitude estimation, commonly, Kalman filtering approaches are applied to solve the problem: the Kalman Filter is applied to solve the drift problem of the inertial signals combing accelerations and angular rates measurements with sensors that can measure inertial quantity without bias. Many examples can be found in the aerospace literature and for four wheeled vehicles. In particular two class of Kalman Filter are proposed:

- *Indirect Kalman Filter*: the indirect formulation of the attitude estimation problem with a KF estimates the errors that act on the inertial signals from the difference between the inertial measurements and aided sensors measurements (*e.g.* [12], [13], [14], [15], [16]);
- *Direct Kalman Filter*: the attitude is directly estimated.

The main drawback of the indirect formulation of a Kalman Filter is that they typically have a low bandwidth, but they are very accurate. The Direct Kalman Filter approach is not widely adopted because it is typically necessary to refer to a dynamic model of the system (relation between forces and accelerations).

In a motorcycle application it is not necessary to obtain very high accuracy, but it is important to have an high bandwidth of the estimator. Thus, it is proposed a novel Direct Kalman Filter that does not require a dynamic description of the process and that has a sufficient bandwidth and accuracy for the application.

In this Book the problem of estimation of the lean angle of a two-wheeled vehicle is addressed in three ways:

- Development of an electro-optical system based on the work reported in [17] and [18];
- Frequency separation based algorithms applied to inertial sensors as in [19], [20] and [21];
- Kalman filtering considering as unique available signals the vehicle speed, three accelerations and three angular rates.

Chapter 3
Preliminaries and basic concepts

In this Chapter some backgrounds that will be useful in the presentation are given. First of all the kinematic of a rigid body will be described (Section 3.1). To illustrate the acquired signals on a vehicle it is always necessary to underline which is the frame to which the measurements are referred to. The dynamic of a vehicle is described by quantities as displacement, velocities and accelerations. All of this variables are represented by vectors. Normally all the available measurements are collected in a vehicle reference frame (Section 3.1.1). The relation between the corresponding variable in the inertial frame is defined by rotational matrix that will be introduced in Section 3.1.2. The definition of the parameters of rotational matrices depends on the chosen angles adopted to represent the attitude (Section 3.1.3). Moreover, also the definition of the angular rates and accelerations of a rigid body depend on the chosen angles (Sections 3.1.4 and 3.1.5).

At the end of the chapter some tools that have been used to solve the estimation and calibration problems are recalled: Kalman filtering (Section 3.2) and Neural Networks (Section 3.3) are briefly introduced.

3.1 Reference kinematics
In this Section some concepts about the kinematic of a rigid body are briefly recalled. The contents of this Section are referred to [22] and [23].

3.1.1 Reference systems definition
The configuration of a body in a three dimensional space is defined by six independent quantities: three of them are related the the translational dynamic while three of them are related to the rotational dynamic. This is consistent with the Chasles' theorem, which state that the displacement of a body may be described by a translation along a line and a rotation about that line. To represent this quantities, vector representation is adopted and this vector has to be defined with respect to a *reference frame* or *coordinate system*. A simple example is reported in Figure 3.1.

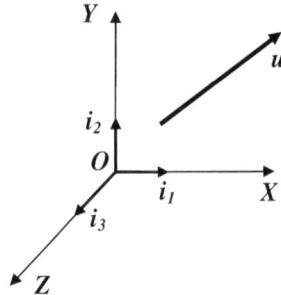

Figure 3.1: Reference frame.

A vector u in a reference system is described by the components of the vector u_1, u_2 and u_3 along the three axis X, Y, Z as

$$u = [u_1 \quad u_2 \quad u_3]^T, \tag{3.1}$$

or as

$$u = u_1\, i_1 + u_2\, i_2 + u_3\, i_3, \tag{3.2}$$

where i_1, i_2 and i_3 are the so called *basis vectors* along the axes X, Y, Z. In the following the representation of a vector as in Equation (3.1) will be adopted.

In the description of the attitude of a body, orthonormal dextral reference frame are considered. A reference frame is said to be *orthogonal* if the basis vector are such that

$$i_1 \cdot i_2 = i_1 \cdot i_3 = i_2 \cdot i_3 = 0. \tag{3.3}$$

If in addiction

$$i_1 \cdot i_1 = i_2 \cdot i_2 = i_3 \cdot i_3 = 1, \tag{3.4}$$

the basis vectors are said to be *orthonormal*. Moreover, if

$$i_1 \times i_2 = i_3 \quad i_2 \times i_3 = i_1 \quad i_3 \times i_1 = i_2, \tag{3.5}$$

the reference frame defined by the basis vectors i_1, i_2 and i_3 is said to be *dextral*.

If a rotation around one of the axes of the frame **XYZ** is applied, then a new ref
has to be defined and the transformation between the two reference frame need

3.1.2 Rotational matrices
In Figure 3.2 the rotation around the Z axis is depicted.

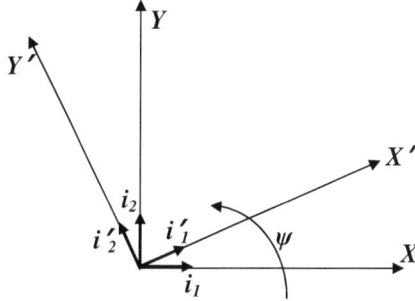

Figure 3.2: Rotation of an angle ψ around the Z axis.

Mathematically this rotation is described by

$$\begin{bmatrix} i_1' \\ i_2' \\ i_3' \end{bmatrix} = \begin{bmatrix} \cos(\psi) & \sin(\psi) & 0 \\ -\sin(\psi) & \cos(\psi) & 0 \\ 0 & 0 & 1 \end{bmatrix} \begin{bmatrix} i_1 \\ i_2 \\ i_3 \end{bmatrix} \tag{3.6a}$$

$$R_Z(\psi) = \begin{bmatrix} \cos(\psi) & \sin(\psi) & 0 \\ -\sin(\psi) & \cos(\psi) & 0 \\ 0 & 0 & 1 \end{bmatrix}. \tag{3.6b}$$

In the context of attitude representation, the cosine matrix $R_Z(\psi)$ is called the *rotation matrix* and it
defines the transformation between two coordinate system [24]. The rotations about the other axes
are defined as

$$R_X(\varphi) = \begin{bmatrix} 1 & 0 & 0 \\ 0 & \cos(\varphi) & \sin(\varphi) \\ 0 & -\sin(\varphi) & \cos(\varphi) \end{bmatrix} \quad R_Y(\vartheta) = \begin{bmatrix} \cos(\vartheta) & 0 & -\sin(\vartheta) \\ 0 & 1 & 0 \\ \sin(\vartheta) & 0 & \cos(\vartheta) \end{bmatrix} \tag{3.7}$$

In the following, for the sake of conciseness, the notation $c_\varepsilon = \cos(\varepsilon)$ and $s_\varepsilon = \sin(\varepsilon)$ is adopted.
The rotation matrix R is an orthogonal matrix, then is satisfies

$$R^T R = I \tag{3.8}$$

I is the identity matrix. From Equation (3.8), it follows that a rotation matrix has to satisfy six constraints of orthonormality and it can have at most three degrees of freedom, so it can be defined by at most three independent parameters.

The composition of successive rotations is computed by matrices multiplication. It has to be noticed that composed rotations are in general non-commutative and that the order of rotation influence the final result (the rotations are commutative just if the adopted rotational axes are parallel). It can be easily verified that if two consecutive rotations of an angle φ_1 and φ_2 are applied around the axis X, then

$$R_X\left(\varphi_1\right)R_X\left(\varphi_2\right) = R_X\left(\varphi_2\right)R_X\left(\varphi_1\right) = R_X\left(\varphi_1 + \varphi_2\right),\tag{3.9}$$

but if rotations around non-parallel axes (*i.e.* rotation around X axis and Y axis) are considered then

$$R_X\left(\varphi\right)R_Y\left(\vartheta\right) \neq R_Y\left(\vartheta\right)R_X\left(\varphi\right).\tag{3.10}$$

Rotations can be composed with a *single frame* method or with a *multi frame* method. In the single frame method all the applied rotations are referred to just one coordinate system (typically the initial reference frame), while in the multi frame method each rotation is referred to the system of coordinates defined by the previous rotation.

In the context of attitude determination, the multi frame method is commonly adopted, then consecutive rotation are composed as follows. Consider the frame **XYZ** as initial frame and the frame *(XYZ)* ″as final frame as depicted in Figure 3.3.

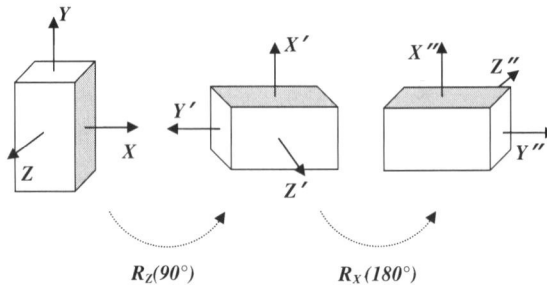

$R_Z(90°)$ $R_X(180°)$

Figure 3.3: Rotation composition in the passive description.

The final frame is defined by

$$\begin{bmatrix} X'' \\ Y'' \\ Z'' \end{bmatrix} = R_X \cdot (180°)R_Z(90°)\begin{bmatrix} X \\ Y \\ Z \end{bmatrix}\tag{3.11}$$

where $R_Z(90)$ represents the rotational matrix around the Z axis of the frame XYX and $R_{X'}(180°)$ is the rotational matrix around the axis X' of the intermediate frame $(XYZ)'$. With the multi frame method to combine consecutive rotations, the rotational matrix i-th is defined in the reference frame $(i$-$1)$-th. By the example shown in Figure 3.3, it can be underlined that the attitude of a coordinate system with respect to a reference frame can be represented by different consecutive rotations. For instance, the coordinate system $(XYZ)''$ can be also be achieved by:

$$\begin{bmatrix} X'' \\ Y'' \\ Z'' \end{bmatrix} = R_Z(90°)R_X(-180°)\begin{bmatrix} X \\ Y \\ Z \end{bmatrix}. \tag{3.12}$$

Two descriptions are in use for the transformation of coordinates: the *passive* description (known also as *alias* description) and the *active* description (known also as *alibi* description). In the passive description, the observer passively adjust his frame while the vector remains immutable. On the other hand, in the active description, the observer is fixed and examines vectors whose direction change in the observer's frame of reference. The two approaches are related by

$$R^{active} = \left(R^{passive}\right)^T. \tag{3.13}$$

In the context of the attitude reconstruction, a passive point of view is in general applied. As we will see in Section 3.1.5, the principal consequence is that the defined rotation matrices transform attitude quantities described in a time invariant reference frame, to the corresponding quantities in a time variant frame.

In the reference literature different sets of parameters are described to define the orientation of a body with respect to a reference frame. The most common and widely used parameters are the Euler angles that will be introduced in the next Section.

3.1.3 Euler angles

Consider three consecutive rotations from the dextral orthonormal coordinate system (XYZ) to the dextral orthonormal coordinate system $(XYZ)'''$ of a body (defined by the basis vectors i_1, i_2 and i_3 fixed on the body). If a multi frame method is taken into account, each rotation is described in the body fixed axes of the considered object and the three angles that define the final rotation are known as *body-referenced Euler angles* [22], [25], [26].

By the adoption of Euler angles to represent the attitude of a body with respect to a reference frame, three degrees of freedom are available. This is in accordance to the available degrees of freedom for a rotational matrix. To constraint the number of free parameters to three it is necessary to impose that the three consecutive rotation have not to be realized around the same axis.

By convention, the Euler angles are defined by the indication of the axis around which the rotation is performed. For instance, if the consecutive performed rotations are $R_Z(\psi)$ then $R_Y(\vartheta)$ and at last $R_X(\varphi)$, then the set of Euler angles are defined as **Z-Y-X**.

Six symmetric sets of Euler angles can be written as

X-Y-X	*X-Z-X*
Y-Z-Y	*Y-X-Y*
Z-X-Z	*Z-Y-Z*

and six asymmetric sets are

X-Y-Z	*X-Z-Y*
Y-Z-X	*Y-X-Z*
Z-X-Y	*Z-Y-X*

The asymmetric sets of Euler angles have been called variously *Cardan angles*, *Tait angles* or *Bryant angles*.

The parameterizations of the rotation matrix in terms of the set **Z-X-Y** is reported in the following equation

$$R_{ZXY}(\varphi, \vartheta, \psi) =$$
$$= \begin{bmatrix} \cos(\vartheta) & 0 & -\sin(\vartheta) \\ 0 & 1 & 0 \\ \sin(\vartheta) & 0 & \cos(\vartheta) \end{bmatrix} \begin{bmatrix} 1 & 0 & 0 \\ 0 & \cos(\varphi) & \sin(\varphi) \\ 0 & -\sin(\varphi) & \cos(\varphi) \end{bmatrix} \begin{bmatrix} \cos(\psi) & \sin(\psi) & 0 \\ -\sin(\psi) & \cos(\psi) & 0 \\ 0 & 0 & 1 \end{bmatrix} =$$
$$= \begin{bmatrix} c_\vartheta c_\psi - s_\varphi s_\vartheta s_\psi & c_\vartheta s_\psi + s_\varphi s_\vartheta c_\psi & -c_\varphi s_\vartheta \\ -c_\varphi s_\psi & c_\varphi c_\psi & s_\varphi \\ s_\vartheta c_\psi + s_\varphi c_\vartheta s_\psi & s_\vartheta s_\psi - s_\varphi c_\vartheta c_\psi & c_\varphi c_\vartheta \end{bmatrix}$$

(3.14)

This parameterization of the attitude of a body is widely used in the automotive context (see [27], [28])

It has to be observed Euler angles are not unique. For instance, for the set **Z-X-Y**

$$R_{ZXY}(\varphi, \vartheta, \psi) = R_{ZXY}(\pi - \varphi, \vartheta - \pi, \psi + \pi).$$

(3.15)

In order to guarantee that the parameters are unique, the following constraints are usually demand

$$-\pi/2 \le \varphi < \pi/2$$
$$0 \le \vartheta < 2\pi$$
$$0 \le \psi < 2\pi$$

(3.16)

This constraints do not represent a limitation in the attitude representation of a vehicle.

All the Euler angle sets are characterized by singularities of the rotation matrix (see [29]). For instance, the set **Z-X-Y** is characterized by singularity when $\varphi \approx \pm\pi/2$. A second aspect that has to be considered when Euler angles are adopted, is how well the set behave if the considered rotations are infinitesimally small. If **Z-X-Y** is taken into account, the rotational matrix associated to infinitesimal angles $(\Delta\varphi, \Delta\vartheta, \Delta\psi)$ has the expression

$$R_{ZXY}(\Delta\varphi, \Delta\vartheta, \Delta\psi) \approx \begin{bmatrix} 1 & \Delta\psi & -\Delta\vartheta \\ -\Delta\psi & 1 & \Delta\varphi \\ \Delta\vartheta & -\Delta\varphi & 1 \end{bmatrix} \tag{3.17}$$

that well behave in the sense that it is a well conditioned matrix.

Even if the Euler angles are characterized by singularity problems, they are widely use because they represent the minimum set of parameters to describe the attitude of a body.

3.1.4 Kinematic relations

The relations between the derivative of the attitude representation and the angular velocity (*kinematic relations*) are given in this Section. The expression of the angular rates of a body depends on the Euler angles that define the transformation between a coordinate system *(XYZ)* and a reference frame *(XYZ)'*. In what follows, the expression of the angular velocity will be firstly derived for a general Euler angles set, then it will be derived the expression for the set **Z-X-Y**.

If the attitude of a body is changing over time, the first derivative of the rotation matrix that represent the changing of the attitude is given by

$$\dot{R}(t) = [[\boldsymbol{\omega}(t)]] R(t) \tag{3.18}$$

where $[[\omega(t)]]$ is an skew matrix in accordance with

$$[[\boldsymbol{\omega}(t)]] = \begin{bmatrix} 0 & -\omega_z(t) & -\omega_y(t) \\ -\omega_z(t) & 0 & \omega_x(t) \\ \omega_y(t) & -\omega_x(t) & 0 \end{bmatrix} \tag{3.19}$$

and $\omega(t) = \begin{bmatrix} \omega_x(t) & \omega_y(t) & \omega_z(t) \end{bmatrix}^T$ is the vector of the *body-reference angular velocity* or simply *angular velocity* defined in the reference frame *(XYZ)'''*.

The vector of the *space-referenced angular velocity* (*i.e.* the angular velocity vector referred to the coordinate system *(XYZ)*) is defined by

$$\Omega(t) = R^T(t) \, \omega(t). \tag{3.20}$$

If the attitude of the body is represented by Euler angles α_1, α_2 and α_3, (α_1 is the first performed rotation and α_3 is the last one) the expression of the body-reference angular velocity vector is given by

$$\begin{bmatrix} \omega_x \\ \omega_y \\ \omega_z \end{bmatrix} = \dot{\alpha}_1 \, \mathbf{n}_1' + \dot{\alpha}_2 \, \mathbf{n}_2' + \dot{\alpha}_3 \, \mathbf{n}_3', \tag{3.21}$$

where the vectors $\mathbf{n}_1'$, $\mathbf{n}_2'$ and $\mathbf{n}_3'$, described in the final coordinate system, represent the axis about which each rotation is performed.

As it has been underlined in the previous Section, the Euler angles are defined with a multi-frame description of the rotation, then, it is necessary to define intermediate rotational matrices that describe the axes around which the rotation is performed in the final coordinate system. The rotation matrix R_{n_i}' is introduced to describe the transformation between the frame in which the rotation around the axes $\mathbf{n}_i$, $i=1,2,3$, is performed and the final coordinate system $(XYZ)'$. Equation (3.21) can be rewritten as

$$\begin{bmatrix} \omega_x \\ \omega_y \\ \omega_z \end{bmatrix} = \dot{\alpha}_1 R_{n_1}' \, \mathbf{n}_1 + \dot{\alpha}_2 R_{n_2}' \, \mathbf{n}_2 + \dot{\alpha}_3 R_{n_3}' \, \mathbf{n}_3 \tag{3.22a}$$

$$\begin{aligned} R_{n_1}' &= R_{n_3}(\alpha_3) R_{n_2}(\alpha_2) \\ R_{n_2}' &= R_{n_3}(\alpha_3) \\ R_{n_3}' &= R_{n_3}(\alpha_3) \end{aligned} \tag{3.22b}$$

where $R_{n_i}(\alpha_i)$ describes the transformation due to the rotation of an angle α_i around the axis $\mathbf{n}_i'$. Then, the angular velocity vector can be expressed as

$$\begin{bmatrix} \omega_x \\ \omega_y \\ \omega_z \end{bmatrix} = R_{n_3}(\alpha_3) S(\alpha_2) \begin{bmatrix} \dot{\alpha}_1 \\ \dot{\alpha}_2 \\ \dot{\alpha}_3 \end{bmatrix} = M(\alpha_1, \alpha_2, \alpha_3) \begin{bmatrix} \dot{\alpha}_1 \\ \dot{\alpha}_2 \\ \dot{\alpha}_3 \end{bmatrix}.$$
$$S(\alpha_2) = \left[R_{n_2}(\alpha_2) n_1 \mid n_2 \mid n_3 \right] \tag{3.23}$$

where $S(\alpha_2)$ is expressed in terms of columns.

The definition of the angular velocity vector can be specified for the case of a **Z-X-Y** set of the Euler angles, thus, the body-referenced angular rates are:

$$\begin{bmatrix} \omega_x \\ \omega_y \\ \omega_z \end{bmatrix} = \begin{bmatrix} c_\vartheta & 0 & -s_\vartheta c_\varphi \\ 0 & 1 & s_\varphi \\ s_\vartheta & 0 & c_\varphi c_\vartheta \end{bmatrix} \begin{bmatrix} \dot{\varphi} \\ \dot{\vartheta} \\ \dot{\psi} \end{bmatrix} = \begin{bmatrix} c_\vartheta \dot{\varphi} - s_\vartheta c_\varphi \dot{\psi} \\ \dot{\vartheta} + s_\varphi \dot{\psi} \\ s_\vartheta \dot{\varphi} + c_\varphi c_\vartheta \dot{\psi} \end{bmatrix} \tag{3.24}$$

To obtain the Euler angular rates, Equation (3.23) can be inverted to give:

$$
\begin{bmatrix} \dot{\alpha}_1 \\ \dot{\alpha}_2 \\ \dot{\alpha}_3 \end{bmatrix} = M^{-1}(\alpha_1,\alpha_2,\alpha_3) \begin{bmatrix} \omega_x \\ \omega_y \\ \omega_z \end{bmatrix}
\tag{3.25}
$$

with

$$
M^{-1}(\alpha_1,\alpha_2,\alpha_3) = S^{-1}(\alpha_2) R_{n_3}^T(\alpha_3)
\tag{3.26}
$$

and it can be shown that for asymmetric set of Euler angles [22]

$$
S^{-1}(\alpha_2) = \frac{1}{\cos(\alpha_2)} \begin{bmatrix} n_1^T \\ \cos(\alpha_2) n_2^T \\ \cos(\alpha_2) n_3^T - \sin(\alpha_2)(n_1 \times n_3)^T \end{bmatrix}.
\tag{3.27}
$$

The relation between the Euler angles and the body-referenced angular velocities is now defined. In the following Section the definition of the acceleration vector of a rigid body is taken into account.

3.1.5 Kinematics accelerations

In this Section the expression of the accelerations of a rigid body is derived as in [23]. As the definition of the angular velocity vector, also the equations that the describe the accelerations of a body depend on the adopted parameterization of the orientation. Again, the parameterization via Euler angles is considered.

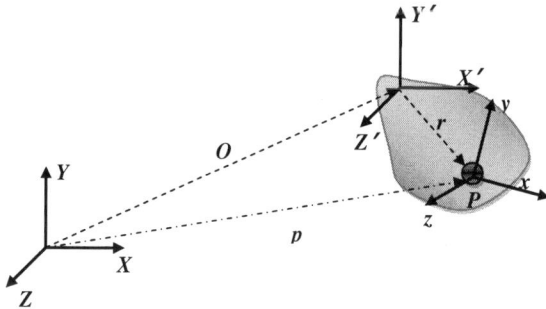

Figure 3.4: Vector composition for the calculation of the kinematic accelerations of a rigid body.

To define the acceleration of a rigid body, it is necessary to introduce three types of coordinate system depicted in Figure 3.4:

- *Global* or *inertial reference frame (XYZ)*: it is a time invariant coordinate system;
- *Body reference frame (xyz)*: it is time variant and fixed on the body, then it translates and rotates with the body;

- *Intermediate reference frames (**XYZ**)*: these are coordinate system that are useful to describe the accelerations of the considered point; normally these frames have the same orientation of the inertial frame.

Consider a point P on a body, as shown in Figure 3.4 the displacement vector p is defined as

$$p_I = O_I + r_I \qquad (3.28)$$

where O is the displacement between the origin of the reference frame and r is the vector that define the position of P in the rigid body with respect the origin of the body coordinate system. The subscript I underlines that the vectors are defined in the inertial frame. Recalling the definition of the rotation matrix with the Euler set **Z-X-Y**, the displacement vector in Equation (3.28) can be rewritten as

$$p_I = O_I + R^T(\varphi, \vartheta, \psi)\, r_b \qquad (3.29)$$

where the subscript b underlines that the vector is defined in the body reference frame.
Differentiating Equation (3.29), the following expression is yield

$$\dot{p}_I = \dot{O}_I + \dot{R}^T(\varphi, \vartheta, \psi)\, r_b + R^T(\varphi, \vartheta, \psi)\, \dot{r}_b \qquad (3.30)$$

where $\dot{p}$ is the velocity vector of the point P and $\dot{O}$ is the velocity vector of the origin of the body reference frame. Substituting Equation (3.18) in Equation (3.30), the following expression is obtained:

$$\dot{p}_I = \dot{O}_I + R^T(\varphi, \vartheta, \psi)\,[[\omega]]r_b + R^T(\varphi, \vartheta, \psi)\, \dot{r}_b \qquad (3.31)$$

where ω is the angular velocity vector referred to the body reference frame.
The expression of the absolute kinematic accelerations of a body can be derived as the derivative of Equation (3.31):

$$\ddot{p}_I = \ddot{O}_I + R^T(\varphi, \vartheta, \psi)\,[[\omega]][[\omega]]r_b + R^T(\varphi, \vartheta, \psi)\,[[\dot{\omega}]]r_b +$$
$$+ 2R^T(\varphi, \vartheta, \psi)\,[[\omega]]\dot{r}_b + R^T(\varphi, \vartheta, \psi)\, \ddot{r}_b \qquad (3.32)$$

In the definition of the kinematic accelerations, the space-reference angular velocity vector is normally less used than the body-reference vector, because the inertia tensor of a body is time variant if described in the time invariant coordinate system, while the body-reference inertia tensor is constant over time.

Finally, the absolute kinematic acceleration can be expressed as a function of the Euler angle rates substituting Equation (3.24) in (3.32) and, applying the coordinate system transformation described by the rotation matrix, the kinematic acceleration in the body reference frame are obtained:

$$\ddot{p}_b = R(\varphi, \vartheta, \psi)\, \ddot{p}_I .$$

(3.33)

The obtained expression of the angular velocities and kinematic acceleration will be useful in the following to describe the measured inertial signals and to develop models for estimation algorithms.

3.2 Kalman Filtering

In this Section the basic equations of discrete Kalman Filter are briefly recalled [30] [31], moreover, the appliance to non-linear systems is discussed through the implementation of the Extended Kalman Filter and the Unscented Kalman Filter.

The discrete-time model for a linear stochastic system has the form

$$x_k = A_{k-1}x_{k-1} + B_k u_k + w_k$$
$$y_k = C_k x_k + D_k u_k + v_k$$

(3.34)

where

- $x_k \in R^n$ is the vector of the state variables;
- $y_k \in R^p$ is the vector of the output variables or measurements;
- $u_k \in R^m$ is the vector of input variables;
- $A_{k-1} \in R^{n \times n}$ is the state transition model which is applied to the previous state x_{k-1};
- $B_k \in R^{n \times m}$ is the control-input model which is applied to the control vector u_k;
- $C_k \in R^{p \times n}$ is the map of the state space into the output space;
- $D_k \in R^{p \times m}$ is the model of the input applied to the output;
- w_k and v_k are process noises which are assumed to have a normal distribution with zero mean and covariance Q_k ($w_k \sim N(0, Q_k)$) and R_k ($v_k \sim N(0, R_k)$) respectively; these noises are assumed to be uncorrelated.

For the system in Equation (3.34), the space variables can be optimally estimated by the application of the *Kalman filter* (KF) which equations are the followings:

$$\hat{x}_k(-) = A_{k-1}x_{k-1}(+) + B_k u_k$$
$$P_k(-) = A_{k-1}P_{k-1}(+)A_{k-1}^T + Q_{k-1}$$
$$K_k = P_k(-)C_k^T \left[C_k P_k(-)C_k^T + R_k \right]^{-1}$$
$$\hat{x}_k(+) = \hat{x}_k(-) + K_k \left[y_k - C_k \hat{x}_k(-) - D_k u_k \right]$$
$$P_k(+) = \left[I - K_k C_k \right] P_k(-)$$

(3.35)

where:

- $\hat{x}_k(-)$ and $\hat{x}_k(+)$ are respectively the a priori and posteriori estimate of the state vector;
- $P_k(-)$ and $P_k(+)$ are respectively the a priori and posteriori estimation of the covariance matrix of the state error estimation $e_k = x_k - \hat{x}_k(+)$;
- K_k is the optimal gain matrix of the Kalman Filter;
- $y_k - C_k\hat{x}_k(-) - D_k u_k$ is the *innovation signal* or *innovation feedback process*.

3.2.1 Extended Kalman Filter

Consider the non-linear discrete system

$$\begin{aligned} x_k &= f(x_{k-1}, u_{k-1}) + w_k \\ y_k &= g(x_k, u_k) + v_k \end{aligned} \qquad (3.36)$$

where w_k and v_k are uncorrelated stochastic noises with a zero mean normal distribution and covariance matrix Q_k ($w_k \sim N(0, Q_k)$) and R_k ($v_k \sim N(0, R_k)$) respectively, $f(\cdot)$ denotes a non-linear transition matrix and $g(\cdot)$ defines the non-linear output transformation matrix of the discrete system. The basic idea of the *Extended Kalman Filter* (EKF) is to linearize the system (3.36) at each time instant around the most recent state estimate $\hat{x}_{k-1}(+)$ and to apply the basic equation of KF (Equation (3.35)) to the obtained linearized model to compute the gain K_k.

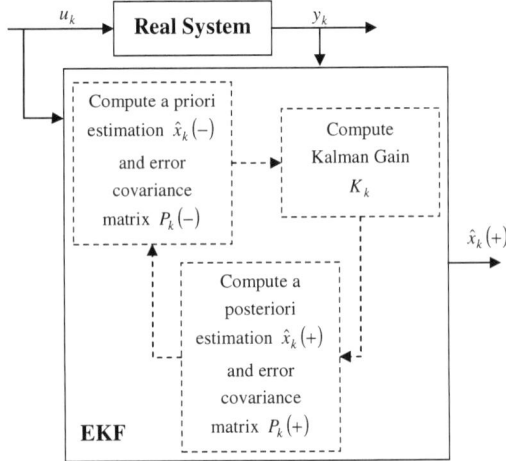

Figure 3.5: Block diagram of Kalman Filtering.

For the system in (3.36), EKF can be defined as follow

$$\hat{x}_k(-) = f\left(\hat{x}_{k-1}(+), u_{k-1}\right)$$
$$\hat{y}_k = g\left(\hat{x}_k(-), u_k\right)$$

$$A(\hat{x}, u, k-1) = \frac{\partial f(x, u)}{\partial x}\bigg|_{x = \hat{x}_{k-1}(+), u = u_{k-1}} \qquad C(\hat{x}, u, k) = \frac{\partial g(x, u)}{\partial x}\bigg|_{x = \hat{x}_k(-), u = u_k}$$

$$\hspace{13cm}(3.37)$$

$$P_k(-) = A(\hat{x}, u, k-1) P_{k-1}(+) A(\hat{x}, u, k-1)^T + Q_{k-1}$$
$$K_k = P_k(-) C(\hat{x}, u, k)^T \left[H_k P_k(-) H_k^T + R_k\right]^{-1}$$
$$\hat{x}_k(+) = \hat{x}_k(-) + K_k\left[y_k - \hat{y}_k\right]$$
$$P_k(+) = \left[I - K_k C(\hat{x}, u, k)\right] P_k(-)$$

As stated before, the matrices that appear in the expression of EKF are obtained evaluating the Jacobian matrix [32] of $f(\cdot)$ in $\left(\hat{x}_{k-1}(+), u_{k-1}\right)$ and $g(\cdot)$ in $\left(\hat{x}_k(-), u_k\right)$. In Figure 3.5, the close loop scheme of the Kalman Filter is shown.

The EKF has been widely applied to estimate the attitude of vehicle in a data-fusion framework. In Chapter 9 the EKF will be applied to estimate the attitude of a motorcycle.

3.2.2 Unscented Kalman Filter

The *Unscented Kalman Filter* (UKF) addressed the approximation of the EKF. As shown in Equation (3.36), in the EKF framework the system state distribution and the noise densities are approximated by *Gaussian Random Variable* (GRV) that are propagated analytically through a first order linearization of the non-linear system (Equation (3.37)). This approximation may introduce large estimation errors and sometimes it may cause divergence of the filter. This drawback can be overcome by using a deterministic sampling approach: the state distribution is still approximated by a GRV, but it is now represented using a minimal set of points [33], [34]. By sampling appropriate points it is possible to better estimate the mean and covariance of the GRV, then, propagating them through the equations of the non-linear system, it is possible to recover a third order approximation (Taylor series expansion) of the mean and covariance of the a posteriori estimation for any linearity. The *Unscented Transformation* (UT) is a method to calculate the statistics o a random variable defined by a non-linear transformation [35], [36] [30]. Consider a random variable $y = f(x)$ and assume that $x \in \Re^L$ has mean $\bar{x}$ and covariance P_x. The statics of y can be calculated considering a matrix $\mathbf{X}$ of $2L + 1$ *sigma* vectors X_i defined by:

$$X_0 = \bar{x}$$
$$X_i = \bar{x} + \left(\sqrt{(L + \lambda)P_x}\right)_i, \quad i = 1, \ldots, L$$
$$X_i = \bar{x} - \left(\sqrt{(L + \lambda)P_x}\right)_i, \quad i = L + 1, \ldots, 2L$$

$$\hspace{13cm}(3.38)$$

where:

- $\lambda = \alpha^2(L + \kappa) - L$ is a scaling parameter;

- α determines the spread of the sigma points around $\bar{x}$ and is usually set to a positive small value (e.g. $1 \le \alpha \le 10^{-4}$);
- κ is a secondary scaling parameter usually set to zero;
- $\left(\sqrt{(L+\lambda)P_x}\right)_i$ is the i-th column of the matrix square root (e.g. lower triangular Cholesky factorization).

The sigma vectors are propagated through the non-linear function:

$$Y_i = f(X_i), \quad i = 1,...,2L \tag{3.39}$$

and the mean and covariance of y are approximated using a weighted sample mean and covariance of the posterior sigma points:

$$
\begin{aligned}
\bar{y} &\approx \sum_{i=0}^{2L} W_i^{(m)} Y_i \\
P_y &= \sum_{i=0}^{2L} W_i^{(c)} (Y_i - \bar{y})(Y_i - \bar{y})^T
\end{aligned}
\tag{3.40}
$$

with weights W_i given by

$$
\begin{aligned}
W_0^{(m)} &= \frac{\lambda}{L+\lambda} \\
W_0^{(c)} &= \frac{\lambda}{L+\lambda} + 1 - \alpha^2 + \beta \\
W_i^{(m)} &= W_i^{(c)} = \frac{1}{2(L+\lambda)}, \quad i = 1,...,2L
\end{aligned}
\tag{3.41}
$$

where β is used to incorporate prior knowledge of the distribution of x (for Gaussian distribution $\beta = 2$ is optimal).

The Unscented Transformation results in approximation that are accurate to the third order for Gaussian inputs for all non-linearity. For non-Gaussian inputs, approximations are accurate to at least second order, with the accuracy of third and higher order statistical moments being determinate by the choice of α and β [30]. A two-dimensional example of the UT is depicted in Figure 3.6: on the left it is shown the true mean and covariance propagation using Monte Carlo sampling; in the middle the propagation due to the linearization realized by EKF is depicted; on the right it is shown that using the UT with just five sigma points it is possible to obtain on accurate approximation of the statistics.

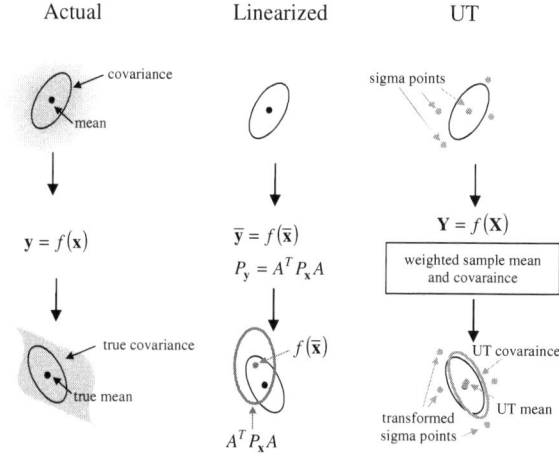

Figure 3.6: Example of the Unscented Transformation for mean and covariance propagation [33], [34].

The Unscented Kalman Filter (UKF) is a straightforward extension of the UT to compute the gain K_k. The state vector is redefined as the concatenation of the original state and noise variables as $x_k^a = \begin{bmatrix} x_k^T & w_k^T & v_k^T \end{bmatrix}^T$. The sigma points selection is applied to the augmented state vector to calculate the corresponding sigma matrix $\mathbf{X}_k^a$. Finally, the UKF equations applied to the system (3.36) are the followings [30].

Initialize with

$$\hat{x}_0^a = \begin{bmatrix} \hat{x}_0^T(+) & 0 & 0 \end{bmatrix}^T$$
$$P_0^a = \begin{bmatrix} P_0(+) & 0 & 0 \\ 0 & Q_0 & 0 \\ 0 & 0 & R_0 \end{bmatrix} \tag{3.42}$$

where $\hat{x}_0(+)$ is the initial value of the state estimation, $P_0(+)$ is the initial value of the covariance of the state error estimation and Q_k and R_k are defined as for the EKF.

For $k \in [1,...,\infty]$, calculate the sigma points:

$$\mathbf{X}_{k-1}^a = \begin{bmatrix} \hat{x}_{k-1}^a & \hat{x}_{k-1}^a + \gamma\sqrt{P_{k-1}^a} & \hat{x}_{k-1}^a - \gamma\sqrt{P_{k-1}^a} \end{bmatrix} \tag{3.43}$$

where $\gamma = \sqrt{L+\lambda}$ and L is the dimension of the state vector x_k^a.

The time update equations are

$$\mathbf{X}^a_{k|k-1} = f\left(\mathbf{X}^x_{k-1}, u_{k-1}, \mathbf{X}^w_{k-1}\right)$$

$$\hat{x}_k(-) = \sum_{i=0}^{2L} W_i^{(m)} \mathbf{X}^x_{i,k|k-1}$$

$$P_k(-) = \sum_{i=0}^{2L} W_i^{(c)} \left(\mathbf{X}^x_{i,k|k-1} - \hat{x}_k(-)\right)\left(\mathbf{X}^x_{i,k|k-1} - \hat{x}_k(-)\right)^T \qquad (3.44)$$

$$\mathbf{Y}_{k|k-1} = g\left(\mathbf{X}^x_{k-1}, u_k, \mathbf{X}^v_{k-1}\right)$$

$$\hat{y}_k = \sum_{i=0}^{2L} W_i^{(m)} \mathbf{Y}_{i,k|k-1}$$

where $\mathbf{X}^a = \left[\left(\mathbf{X}^x\right)^T \ \left(\mathbf{X}^w\right)^T \ \left(\mathbf{X}^v\right)^T\right]^T$, and the measurement-update equations are

$$P_{y_k y_k} = \sum_{i=0}^{2L} W_i^{(c)} \left(\mathbf{Y}_{i,k|k-1} - \hat{y}_k\right)\left(\mathbf{Y}_{i,k|k-1} - \hat{y}_k\right)^T$$

$$P_{x_k y_k} = \sum_{i=0}^{2L} W_i^{(c)} \left(\mathbf{X}^x_{i,k|k-1} - \hat{x}_k(-)\right)\left(\mathbf{Y}_{i,k|k-1} - \hat{y}_k^-\right)^T$$

$$K_k = P_{x_k y_k} P_{y_k y_k}^{-1} \qquad (3.45)$$

$$\hat{x}_k(+) = \hat{x}_k(-) + K_k\left[y_k - \hat{y}_k\right]$$

$$P_k(+) = P_k(-) - K_k P_{y_k y_k} K_k^T$$

In the previous equations W_i, λ and L are defined by the UT and $\mathbf{X}^a_i$, $\mathbf{Y}_i$ represent the i-th column of the matrix $\mathbf{X}^a_i$ and $\mathbf{Y}_i$ respectively.

In Chapter 9, also the UKF will be plied to the problem of attitude estimation with inertial sensors in motorcycle applications and the estimation performance will be compared to the results of the KF highlighting the principal advantages of the unscented transformation.

In the next subsection some basic concepts on Artificial Neural Network are introduced.

3.3 Artificial Neural Networks

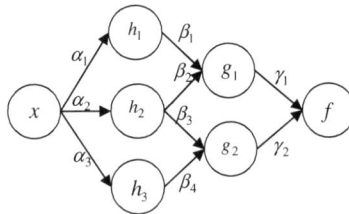

Figure 3.7: Feed forward Neural Network structure.

Some basic concepts on Neural Network are now given [30], [37], [38], [39], [40], [41]. An Artificial Neural Network (ANN), usually called Neural Network (NN), is a non-linear statistical data modeling tool that can be used to model complex relation between input and output quantities.

Basically, a NN involves a network of simple processing elements called *neurons* which can exhibit a complex global behavior. The output of the network depends on connections between neurons and parameters that define them.

Artificial Neural Networks essentially define a non-linear function f between inputs x and outputs y ($f : X \to Y$). As a network, the function $f(x)$ is defined as a non-linear weighted sum of other functions $g_i(x)$ as

$$y = f(x) = K\left(\sum_i \gamma_i g_i(x)\right) \tag{3.46}$$

where $g_i(x)$ can be further defined as a composition of other functions and $K(\cdot)$ is some predefined function, such as hyperbolic tangent.

To represent the composition of function by which $f(x)$ is defined, a graphical paradigm is commonly used as depicted in Figure 3.7, in which each node represents a predefined function and the weights are highlighted on the arcs.

If the graph that describes the NN is a directs acyclic graph, the networks is called feed-forward (Figure 3.7).

Basically the function $f(x)$ is defined by a learning procedure that means that given a set of functions F, using a set of observations $\bar{x} \in X$ to find $f^0 \in F$ that is optimal in some sense. Therefore, a functional cost $J : F \times X \to \Re$ has to be defined, such that $J(f^0, \bar{x}) \leq J(f, \bar{x}) \forall\ f \in F$. The definition of the functional cost depends on the task that has to be solved. If a set of pairs $(\bar{x}, \bar{y}), \bar{x} \in X, \bar{y} \in Y$ is given and the aim of the task is to identify $f : X \to Y$ in the allowed class of function, then the supervised learning paradigm can be applied. In this framework, the mean squared error

$$J(f, \bar{x}) = E\left[(\bar{y} - f(\bar{x}))^2\right] \tag{3.47}$$

is commonly adopted, where $E[\cdot]$ is the well known expected value operator.

In this thesis two classes of NN are considered: *Feed-forward Neural Network* and *Radial Basis Network*.

3.3.1 Feed-forward Neural Network

As stated before, for this class of NN the graph does not show a direct cycle, then the information flow forward from the input layer, through the hidden layers, to the output layer.

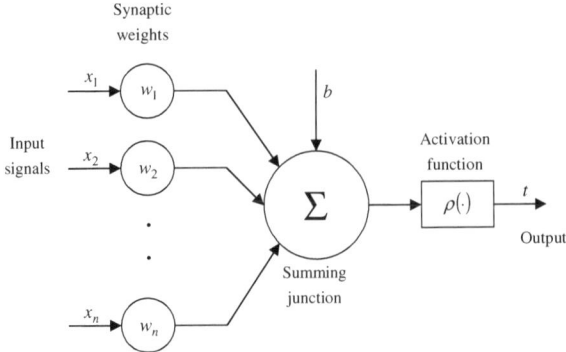

Figure 3.8: Model of a perceptron.

In this context, the neuron is a *perceptron* depicted in Figure 3.8. In mathematical terms, the neuron output is defined by

$$t = \sigma\left(\sum_{i=1}^{n} w_i x_i + b\right) \tag{3.48}$$

where x_i, $i = 1,...,n$, are the input signals, w_i, $i = 1,...,n$, are the synaptic weights of the neuron, b is the bias, ρ is the activation function and t is the output signal of the neuron.

Due to the differentiability and symmetric properties, the adopted activation function is the *symmetric sigmoid function* recalled in the following Equation:

$$\rho(x) = \frac{1 - e^{-x}}{1 + e^{-x}} \tag{3.49}$$

Recalling the *Universal Approximation Theorem* [42], it is often considered a multilayer feed-forward network with a single hidden layer that contains a finite number N of neurons (perceptrons) and a linear output layer (weighted sum of the outputs of the neurons in the hidden layer). Then, for each output signal y of the NN, the number of parameters of a feed-forward NN is equal to $(n + 2)N$. The learning procedure can be performed by the application of the well known back-propagation algorithm: given the observations $(\bar{x}, \bar{y})$, $\bar{x} \in X$, $\bar{y} \in Y$ and the number N of the neurons in the hidden layer, the error between the output of the network $f(\bar{x})$ and the correct answer $\bar{y}$ is computed, and the error is fed back into the network for adjusting the weights in order to reduce the functional cost; this procedure is repeated until convergence.

The problem of definition of the complexity N of the hidden layer is not trivial. To solve this problem and to avoid overfitting [43] various methods can be adopted:

- *Cross validation* [44], [45]
- *Training with penalty terms* [46]

- *Weight decay* and *node pruning* [47], [48]
- *Network Information Criterion* [49]
- *Heuristic methods*

Moreover, the problem of local minima has to be taken into account. This problem can be overcome considering different initial conditions for each degree of complexity of the network, as a consequence the optimal network for the considered number on neurons in the hidden layer is the trained network for which the best performance is achieved.

3.3.2 Radial Basis Network

Referring to Figure 3.8, in a Radial Basis Network (RBN) the activation function is a Radial Basis Function (RBF) that is a Gaussian function

$$\rho(x) = e^{-\sigma\|x\|^2} \qquad (3.50)$$

where the norm is typically the Euclidean distance.

In this framework the bias b (Figure 3.8) is called center of the Gaussian distribution and σ is the spread of the distribution.

The RBNs are structured with a single hidden layer with non-linear RBF and a linear output layer, hence, as the feed-forward NN, also this class of NNs are universal approximators.

The center of the Gaussian distribution is normally different for each neuron, while the spread is common; as a consequence, for each output signal y of the network the parameters of the network are the inputs and outputs weights, the centers of the neurons and the value of the spread.

A big advantage of the RBF network respect to the feed-forward neural network is that if the centers of the neurons and the value of the spread are fixed, then the weights can be obtained solving a linear problem that brings a unique local minimum of the functional cost defined in Equation (3.47) [38].

For a chosen number N of neurons in the hidden layer and for a fixed value of the width of the Gaussian function, the training of the RBN is performed in two phases:

1. The centers are fixed by
    - Sampling the input instances
    - Applying Orthogonal Least Square Learning Algorithm
    - Clustering the observations and choosing the mean of the clusters as centers
2. The weights are computed minimizing the functional cost and solving a linear problem.

The complexity of the RBN depends both of the number of neurons and on the value of σ. In particular, the larger is N, the smaller is the σ for which the best performance is achieved. Finally, the complexity can be analyzed with the same methods described in Section 3.3.1.

Chapter 4
Roll angle estimation: preliminaries

This Chapter is devoted to the introduction of the attitude parameters of a motorcycle and to the description of the environments that have been used to test the estimation algorithms presented in the followings Chapter.

Basically, in the attitude estimation context, the set of Euler attitude angles that describe the orientation of the body reference coordinate system (body reference moving frame) with respect to the inertial reference system (fixed frame) are considered to be the attitude angles of the vehicle. In many control problems that arise in the automotive context, the design of the controller requires the knowledge of the inclination of the motorbike with respect to the road (*road attitude angles*). The tire contact forces are defined as a function of the road lean angle on the vehicle (see [5], [1]), thus, in many motorcycle control problems (see [50], [51], [2], [6]) such as traction control [52], [53], stability control [54], [55], [56] and brake control [57], [9] of a motorcycle, the *road roll angle* is the fundamental parameter that is necessary for a parameterization of the controller. In Section 4.1 the difference between the Euler attitude angle and the road attitude angle will be studied.

Roughly speaking, the lean angle of a motorcycle can be defined as the inclination φ_I (*inertial roll angle*) that is necessary to reach the equilibrium of the moments when the vehicle is leaned over. In Section 4.2, the difference between the Euler attitude angle and the inertial one will be deeply analyzed.

The end of the Chapter (Section 4.3) is devoted to the presentation of the simulation environment and the experimental set-up.

4.1 Euler and road attitude angles
The attitude of a rigid body can be represented in different ways. In [22] a survey on the attitude representation is reported. In this application, the motorcycle attitude is described by the Euler angles (Section 3.1.3) even if this choice brings to a non-linear description of the vehicle kinematic.

To define the attitude of the motorcycle is first of all necessary to give a definition to the vehicle's rotation axes (Figure 4.1) [3], [58]. The roll axis is the axis of the ground around which the motorcycle can roll, remembering that the attitude does not depend on the position of the rotational axis, without loss of generality the roll axis can be considered as the longitudinal axis of the motorcycle. The pitch axis is the axis of rotation due to the lowering of the motorcycle's steering head when it is turned. The yaw axis is the axis of vertical rotation of the motorcycle. Conventionally, the *roll* angle is indicated with φ, the *pitch* angle with ϑ and the *yaw* angle with ψ.

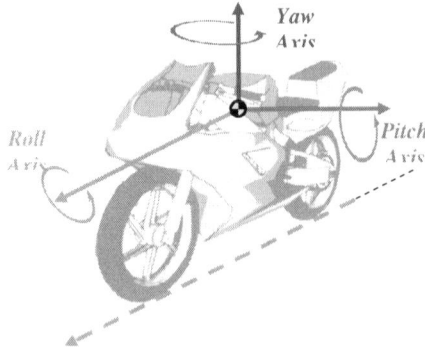

Figure 4.1: Rotation axes definition.

Once the rotation axes have been defined, the *body reference frame* (**xyz**) of the vehicle is chosen as a dextral time variant coordinate system positioned on the COG (Center Of Gravity) of the vehicle, with the x axis in accordance to the roll axis of the vehicle and the z axis in accordance to the yaw axis of the vehicle.

The orientation of the body reference frame can be described with respect to the *inertial reference frame* (**XYZ**) that is a dextral fixed time invariant coordinate system. In this case the attitude angles of the vehicle are named *absolute Euler attitude angles* (or simply Euler attitude angles) and they will be indicated with the subscript E.

The attitude of the vehicle can be also defined referring to the *road reference frame* that is a that is a dextral fixed coordinate system which vertical axis is always perpendicular to the road plane and the yaw rotation with respect to the inertial reference frame is null. In this case the attitude angles are named *road attitude angles* and they will be indicated with the subscript R.

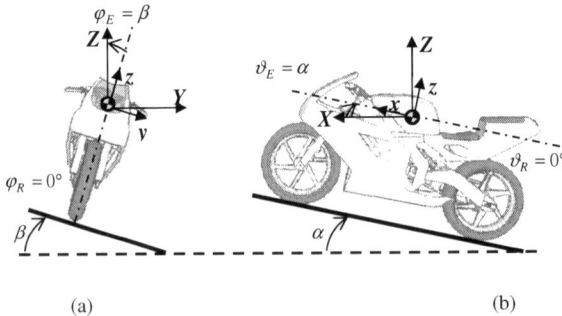

(a)   (b)

Figure 4.2: Road and inertial attitude angles.

The difference between the Euler and road attitude angles is due to the presence of slopes and banks on the track that are interpreted as a pitch and roll dynamic respectively (Figure 4.2). Referring to the slope with α and to the bank with β, the inertial and road attitude angles are related by

$$\varphi_E = \varphi_R + \beta$$
$$\vartheta_E = \vartheta_R + \alpha .$$
$$\psi_E = \psi_R$$

(4.1)

The principal drawback of the inertial measurements is that they are informative about the Euler attitude angles and it is not trivial to split the road inclinations from the road attitude angles. This aspect will be underlined in Chapter 7 developing the models of the inertial measurements.

4.2 Inertial roll angle

A motorcycle in a turning condition is subject to a moments balance around the tires point of contact that defines the lean angle of the vehicle.

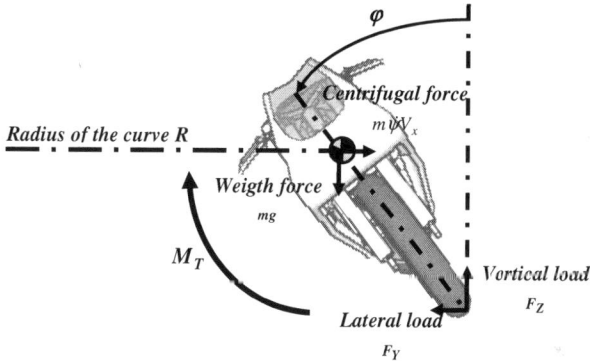

Figure 4.3: The force balance on a curve.

First of all consider that the motorcycle runs along a turn of constant radius at a constant speed (steady state condition), thus the gyroscope effect [3] is negligible. Moreover, making the following simplifying hypothesis:

- The track is in plane;
- The tires thickness is null;
- The COG of the rider belong to the principal symmetry plane of the motorcycle;

a motorcycle in a turning condition is subject to (Figure 4.3):

- Restoring moment due to the centrifugal force $m\dot{\psi}V_x$
- Tilting moment due to the weight force mg .

where $\dot{\psi}$ is the yaw rate, V_x is the speed of the vehicle and m is the mass of the motorcycle.

Thus, in a ideal turning condition the roll moments equilibrium equation is

$$mgHs_{\varphi} = m\dot{\psi}V_x Hc_{\varphi} \qquad (4.2)$$

where H is the height of the COG the vehicle. From Equation (4.2)

$$t_{\varphi} = \frac{\dot{\psi}V_x}{g}, \qquad (4.3)$$

where $t_{\varphi} = \tan(\varphi)$, thus, the *inertial roll angle* is defined by

$$\varphi_i = \arctan\left(\frac{\dot{\psi}V_x}{g}\right). \qquad (4.4)$$

The moments balance can be analyzed more deeply.

First of all Equation (4.4) is valid just in a steady state turning condition. In real application, the steady state turning condition is just an idealization of the motorcycle dynamic, as a consequence, estimation errors occur due to the transversal dynamics (*i.e.* longitudinal dynamic, heave dynamic) and pitch dynamic.

An error is introduced also by the slope of the track, and, even if a steady state turning condition is satisfied and the track is in plane, the foremost approximations are due to the effect of the tires thickness and the displacement of the COG due to the rider movements. In fact, the inertial lean angle is the inclination of the axis passing through the tire POC and the COG of the system rider-motorcycle; as a consequence, it will strongly depend on the position of the POC and the COG of the system.

4.2.1 Effect of the tires thickness on the inertial roll angle

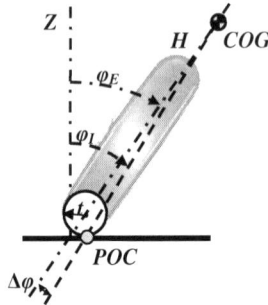

Figure 4.4: Roll angle of a motorcycle with real tires.

Ideally, the resultant of the centrifugal and weight force pass through the tires contact points on the road plane. Moreover, if the thickness of the tires is null and the steering angle of the motorbike is small, the resultant lies in the motorcycle longitudinal plane of symmetry. In a more real condition the contact points of the tires do not lies in the motorcycle longitudinal plane of symmetry, they are laterally misaligned (Figure 4.4) and the roll angle φ_E of the motorbike differs from the inertial roll angle φ_I defined by the equilibrium of the moments as in Equation (4.4), in fact

$$\varphi_E = \varphi_I + \Delta\varphi. \tag{4.5}$$

Consider that the tires of the vehicle have a thickness equal to $2t$, thus the difference $\Delta\varphi$ between the Euler roll angle and the inertial lean angle is defined by

$$\Delta\varphi = \frac{t}{H-t}\varphi_I, \tag{4.6}$$

as a consequence, the Euler roll angle in a steady state turning condition is defined by

$$\varphi_E = \left(1 + \frac{t}{H-t}\right)\arctan\left(\frac{\dot{\psi}V_x}{g}\right). \tag{4.7}$$

Consider $t=5$ cm and $H=50$ cm (these are typical values for an high performance motorbike): the difference $\Delta\varphi$ the ESR introduced by (4.6) is proportional to $t/(H-t) \cong 10\%$.

4.2.2 Effect of the rider movements on the inertial roll angle

When a motorcycle is in a turning condition, the rider tends to move its own COG outside of the plane of symmetry of the vehicle. The corresponding effect is a displacement of the COG of the system rider-motorcycle.

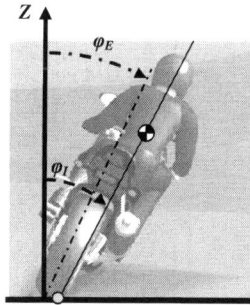

Figure 4.5: Effect of the COG displacement on the inertial roll angle.

The error introduced by the displacement of the COG strongly depends on the driving style of the rider. What is interesting is the relation with error introduced by the non-null tires thickness. The rider tends to move the COG of the system inside of the curve as depicted in Figure 4.5. As a consequence, the displacement of the COG tends to compensate the effect of the displacement of the tires POC due to a non-null thickness.

4.3 Simulation environment and experimental platform

The results presented form Chapter 6 to Chapter 9 are based both on the data obtained on the MSC BikeSim® multibody simulator for two-wheeled vehicles and on experimental data collected with an Aprilia Tuono1000 Factory. Due to confidentiality, not many details will given about the experimental set-up.

In the first part of this Section the simulator is briefly introduced, in the second part the experimental setup will be detailed.

4.3.1 BikeSim

MSC BikeSim® [59] is a full-fledged motorbike multibody simulator [60], [61], [62] whose mathematical model is based on the work described in [28]. In Figure 4.6a picture of the simulator is depicted. Also other multibody simulators such as the one presented in [63], [64], [65] have been partially used for preliminary analysis.

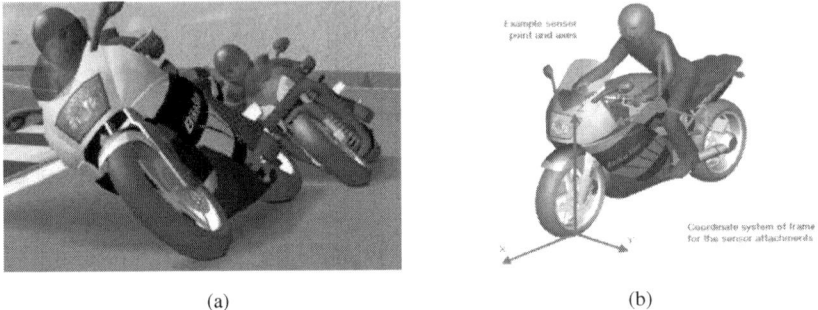

(a) (b)

Figure 4.6: (a) Picture of the simulator environment. (b) Positioning of the virtual IMU.

The motorcycle adopted for the simulation in the MSC BikeSim® can be equipped with different virtual inertial sensors. The interesting signals are the ones collected positioning the virtual IMU (*Inertial Measurement Unit*) in the center of mass of the vehicle (see Figure 4.6b). It is obvious that in a real application is not possible to mount an IMU in the center of mass of the vehicle, but the approximation introduced is negligible.

The interesting signals that are provided by the simulator are:

- Euler angles ZXY (see Chapter 3)
- Inclination angles of the road
- Absolute angular rates and angular rates measured on the motorcycle
- Absolute accelerations and accelerations measured on the motorcycle
- Velocity of the vehicle.

These quantities are useful to validate the designed algorithms in order to verify in a noiseless condition which are the limits of the performance.

4.3.2 Experimental set-up

The analysis presented in this work has been supported with experimental data. The experimental data have been collected with an Aprilia TUONO1000 Factory shown in Figure 4.7. On the test vehicle, the following set of sensors was employed: one Inertial Measurement Unit (IMU) composed by three 1-axis Silicon Sensing MEMS (Micro Electro-Mechanical Systems) gyroscopes (CRS-07) and a 3-axis ST-Microelettronics MEMS accelerometer (LIS3L02AS4); two Hall-effect wheel encoders with 48 teeth to measure the front and rear wheel rotational speed. It is well known that the electro-optical sensors represents the unique technology to directly measure the attitude of the motorbike with respect to the asphalt [18], [17]. Therefore, the reference roll angle is calculated via electro-optical measurement as described in Chapter 5.

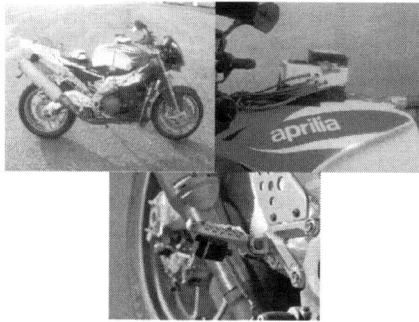

Figure 4.7: Instrumented Aprilia TUONO1000 factory with inertial measuring unit and electro-optical triangulator.

The inertial platform has been mounted on the fuel tank of the vehicle that is not far from the COG, while the electro-optical sensors have been fixed under the pedals of the motorcycle with a distance between them of 40 cm.

Figure 4.8: Torpedo ECU developed by E-Shock.

The signals are acquired on the ECU depicted in Figure 4.8 with a sampling frequency of 100 Hz and filtered with a second order filter at a cut-off frequency of 7 Hz.

The experimental test have been conducted on the circuit Santa Monica in Misano Adriatico (see Figure 4.9a), on the circuit Enzo Ferrari in Imola (see Figure 4.9b) and on the Mugello Circuit (see Figure 4.9c). The characteristics of these tracks are:

- *Santa Monica circuit*: length is 4226 m, the corner radius range is 20 m-120 m and the maximum slope is 1.5% (0.8°).

- *Enzo Ferrari circuit*: length is 4959 m, the corner radius range is 11 m-60 m, the maximum slope up is 7.81% (4.46°) and the maximum slope down is 6.22% (3.56°)
- *Mugello Circuit*: length is 4959 m.

In the followings Chapters, the it will be described the estimation of the lean angle of a two wheeled vehicle with electro-optical measurements, GPS signals and inertial signals (accelerations and angular rates).

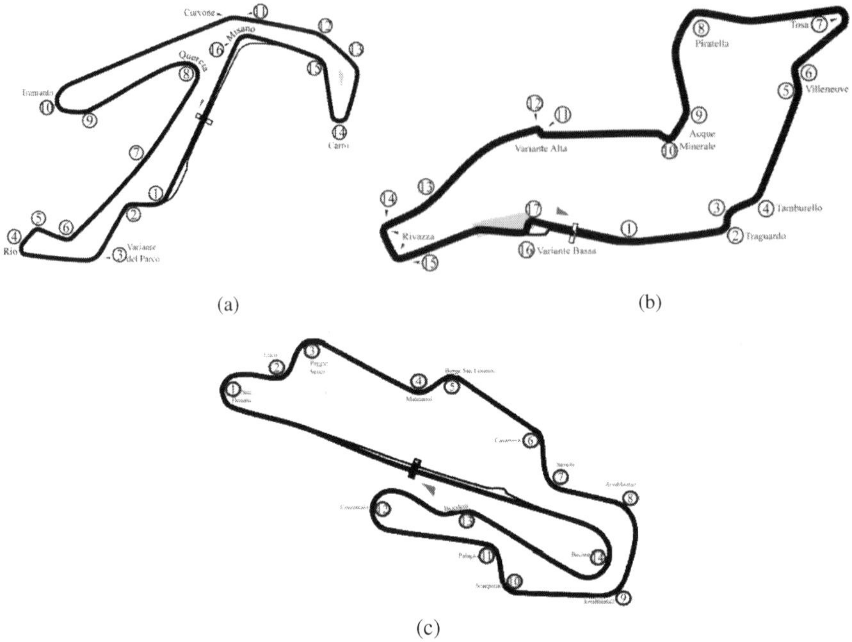

(a)

(b)

(c)

Figure 4.9: (a) Santa Monica Circuit in Misano Adriatico; (b) Enzo Ferrari Circuit in Imola; (c) Mugello Circuit.

Chapter 5
Optical measurement of the attitude of a motorcycle

The aim of this Chapter is to describe the design and characterization of a novel optical sensor well suitable, both from a technological and economical point of view, for real-time tilt measurements in hypersport motorcycles for racing applications. As a matter of fact, motorcycles are a very demanding application for measurements systems, because of significant vibrations due to the engine, high vehicle speeds, and different asphalt and weather conditions in which they happen to operate.

The electro-optical measurement system is the only way to obtain a lean angle measurement such that:

- The measured lean angle is referred to the road;
- The accuracy is high;
- The signal can be acquired in real time;
- The algorithmic effort is low.

Due to these characteristics, the proposed system will be adopted in the following Chapters to construct the refence signal to which compare the lean angle estimated through position and inertial signals.

The structure of the Chapter is as follows. Primarily, in Section 5.1 the triangulation principle for distance measurement is introduced. Section 5.2 is devoted to the problem statements, introducing the notation required and showing the main difficulties to be handled in order to design a reliable tilt angle sensor for the considered application. In Section 5.3 the sensor design and characterization are addressed. Section 5.4 provides a discussion on the sources of measurements uncertainty and describes the optimization of the mounting parameter values to minimize both measurement uncertainty and maximum error. Section 5.6 presents an analysis to show that the electro-optical measurement system is not influenced b the tires thickness. Finally, Section 5.7 assesses the performance of the designed sensor presenting experimental results obtained on a test track. The Chapter is ended with concluding remarks and an outlook to future work.

5.1 Laser telemeters

A laser telemeter is an instrument to measure the distance of a remote target. There are basically three measurement principles [66]:

- *Triangulation*: consider two points separated by a known distance D placed perpendicular to the line of sight; by measuring the angle δ formed by the two lines of sight, the distance is found as $L \cong D/\delta$.

- *Time of flight*: a light beam is propagated form a laser source to a target and back; by measuring the time delay $T = 2L/c$, where c the speed of light, the distance follows as $L = cT/2$.

- *Interferometry*: a laser beam is propagated to the target; the returned field is detected coherently by beating with a reference field on the photodetector and a signal of the form $\cos(2ks)$, where $k = 2\pi/\lambda$ and λ is the wave length, and the distance is measured as increments in unit $\lambda/2$.

Triangulation is the easiest to be implemented and it is widely used in application for distance measuring between the vehicle chassis and the road. Therefore, in what fallows the triangulation principles is going to be briefly described.

5.1.1 Triangulation principle

To describe the triangulation principle the basic scheme in Figure 5.1 in which passive triangulation is considered (the target is considered to be self luminous).

The rotatable mirror and the beamsplitter permit to superimpose the image seen at points **a** and **b**. Knowing the angle δ between the orientation of the beamsplitter and the mirror and the mounting distance D, the distance L is defined by

$$L = \frac{D}{\tan(\delta)} \cong \frac{D}{\delta}. \tag{5.1}$$

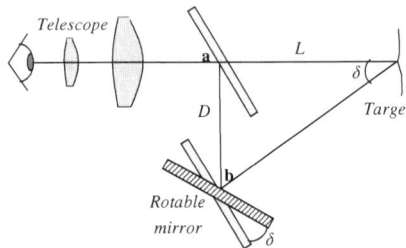

Figure 5.1: Basic scheme of a triangulation measurement.

A source of light (*emitter*) can be added to aim the target and a Position Sensitive Detector (PSD) can be adopted to sense the returned light (*receiver*).

The scheme depicted in Figure 5.2 is commonly adopted to remove moving parts and get a faster response. This is the layout of an active telemeter for short range measurement (from 0.1 m to 10 m) that uses a semiconductor laser and a PSD. The distance L is yield by

$$\frac{D}{L} = \frac{x}{f} = \tan(\delta)$$

$$L = \frac{Df}{x}$$

(5.2)

where x is the position of the spot of light on the PSD and f is the distance between the IF (Interference Filter) and the PSD.

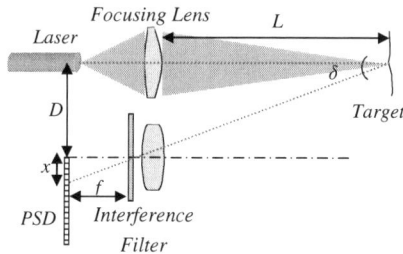

Figure 5.2: Scheme of a triangulation telemeter with active illumination and static readout.

The distance range of the triangulation telemeter depends on the parameters D and f and on the ratio between the emitted and received optical power. The received optical power is a function of $1/L^2$: the maximum measurable distance L is defined by the condition at which the received power is less than the dark-noise of the PSD (optical power due to the ambient light).

5.1.2 Main sources of errors in triangulation telemeters
The measurement of a triangulation telemeters is affected by various sources of error that can classified as *optical errors* and *electronic errors*.

Some of the foremost optical errors are

- *Misalignment*: errors due to the mounting uncertainty; the most critical misalignments are due to the beam of the laser not perpendicular to PSD and the PSD not parallel to the IF;
- *Optical frequency drift*: the optical frequency of the laser is a function of the temperature and it is critical if the wavelength of the laser is out of the IF bandwidth because the distance measurement range of the sensor dramatically reduces.

The most critical electronic errors are:

- *Temperature drift*: the dark noise of the PSD depends on the temperature of the device;
- *Analog to Digital resolution*: the LSB (Least Significant Bit) of the converter limits the distance resolution of the device; the output voltage is typically proportional to the value of x, as a consequence, it is proportional to $1/L$ and there is a non linear relation between the resolution of the ADC and the equivalent distance resolution.

In what follows, an electro-optical measurement system for roll angle measuring is described. The adopted sensor is a triangulation telemeter with a low voltage resolution. It will be shown that the low resolution is not critical for lean angle measuring and how the problem of interaction between the received optical power and the noise due to solar light can be overcome.

5.2 Problem statement

A geometrical description of the system for measuring the tilt angle of a motorbike is depicted in Figure 5.3. In the simplest configuration, two triangulation telemeters – one for each side of the motorcycle– are used. From the figure it is possible to note that when the motorbike axis is not vertical, the distances d_1 and d_2 between the marked points and the ground become unequal.

The sensors shown in Figure 5.3 need to be symmetrically located with respect to the bike axis and must lay on a horizontal line when the bike axis is vertical.

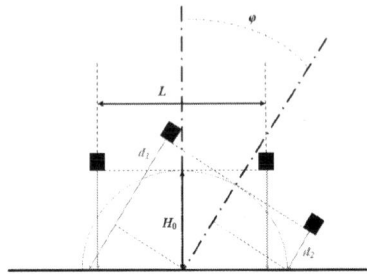

Figure 5.3: Definition of the tilt angle φ between the motorbike axis and the vertical direction. Filled squares indicate the distance sensors. L: lateral displacement between the two distance sensors; H_0: distance between each sensor and the ground when the bike axis is vertical; d_1 and d_2: distances between the marked points and the ground.

From the distance difference (d_1-d_2) the tilt angle is readily obtainable as

$$\varphi_R = \arctan\left(\frac{d_1 - d_2}{L}\right). \tag{5.3}$$

The angle φ is the tilt angle of the motorcycle measured with respect to the road referred here after as road roll angle (φ_R).

The measured distances d_1 and d_2 can be expressed as functions of the mounting parameters L and H_0 (see Figure 5.3) as

$$\begin{cases} d_1 = H_0 + \dfrac{L}{2}\tan(\varphi_R) \\ d_2 = H_0 - \dfrac{L}{2}\tan(\varphi_R) \end{cases} \tag{5.4}$$

Therefore, in order to measure the tilt angle in real time, one can use two distance sensors capable of real-time measurements of the two distances from the asphalt and a straightforward calculation.

For such sensing application, different kinds of contact-less electro-optical technology can be employed. The optical telemeter, based on time–of–flight measurement, is not a good candidate for this application, due to the excessive size, weight and cost [66]. Other techniques exploit the coherence property of the laser light [67], [66], but they are really difficult to implement on a target moving at high speed, such as the asphalt. The more viable solution for this specific application could be an optical triangulator [18], commercially available at very low cost. The output of such optical distance sensors can be extremely accurate and rapid enough for this specific application.

The optimal values of the mounting parameters L and H_0 (see Figure 5.3) depend on the metrological characteristics of the specific optical sensors adopted and, in this work, they are chosen to minimize the measurement uncertainty around zero tilt angle, according to the constraints due to both the measuring range of the sensors and the available space for physical mounting on the motorbike. The analysis of the measurement uncertainty will be thoroughly discussed in Section 5.4. Obviously, the weight and size of the optical sensors need to be kept as small as possible whereas the measurement reliability under different ambient conditions is a main issue as well. Figure 5.4 shows some possible mounting locations of the optical sensors on the motorbike.

Figure 5.4: Possible locations of distance sensors on the motorbike (left) and photograph of the mounted sensor (right).

5.3 Design of an optical triangulator for motorcycle application

In the first prototype version of the optical sensor (see [18]), a commercial triangulator was used, characterized by a measurement range of 15-100 cm and an output voltage in the range 0.5-2.5 V (quantization $\Delta V = 20$ mV), non-linearly related to the measured distance. In this sensor, the emitter was a LED at wavelength $\lambda \cong 900$ nm with 0.1 mW average output power and 1 kHz amplitude modulation (on-off, with 10 % duty-cycle) of the optical signal. The 1 kHz modulation of the optical signal is used to reduce interference from quasi-DC external light sources and significantly increases the signal-to-noise ratio at the detector with respect to using DC signals. The PSD used to receive the back-diffused light from the ground surface was protected by a colored plastic cover, providing for some filtering of visible background radiation. Distance data, in the form of analog voltage values, are available at the sensor output with a refresh frequency of 25 Hz.

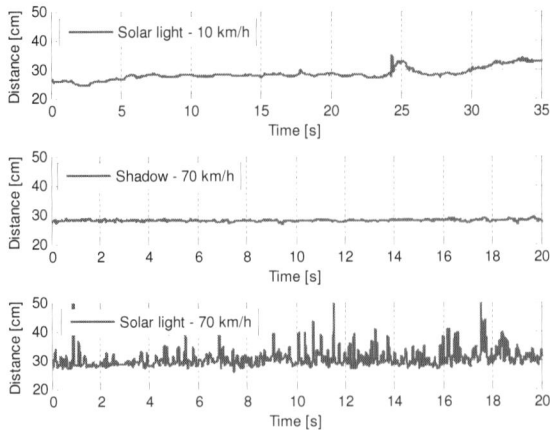

Figure 5.5: Time histories of the measured distance obtained with the LED emitter: solar lighting and speed of 10 km/h (top), shadow and speed of 70 km/h (middle), solar lighting and speed of 70 km/h (bottom).

After obtaining the sensor characteristic and mounting two sensors on the motorbike, several tests were performed to ascertain how the vehicle speed could affect the output signal of the distance sensor and if there might be other significant noise sources coming from the external environment. To better isolate the problem, the tests were all taken in the same predefined straight-line freeway path, which was covered at different constant speeds, in both directions. Thanks to this procedure, in a round trip of the path each of the two sensors can be tested once under solar lighting exposure and once in the vehicle shade. Note that all these tests should, in principle, provide approximately the same values of measured distances from ground, for both sensors and for both directions along the path. In practice, due to non-perfect repeatability of the road track and slight motorbike inclination, the measured distances are in the range from 25 cm to 35 cm.

Figure 5.5 shows the time histories of the distance signals measured by the right-side sensor at two different speed values of 10 km/h and 70 km/h, respectively. The results in the top plot show that the condition of low speed and solar lighting does not affect the S/N ratio of the measured distance, thus maintaining a very good accuracy level. This also happens with the combination of high speed and shadowing (middle plot in Figure 5.5). However, significant problems appear at a motorcycle speed of 70 km/h in the presence of solar lighting: in fact (see the bottom plot in Figure 5.5), in this condition the distance measurements are clearly affected by a significant noise contribution which makes the sensor output highly unreliable.

Considering the results of this experimental analysis, it can conclude that a significant optical interference phenomenon exists, which appears to be due to a combination of high vehicle speed and strong solar lighting coming as diffused light from the asphalt. In addition, it can be observed that the interference intensity significantly depends on the speed value and asphalt surface roughness. This was expected, since the spatial periodicity introduced by the asphalt roughness is converted into a frequency distribution of detected solar interference and the Fourier frequencies of this last spectrum directly depend on the speed at which the optical triangulator (and hence the motorbike) is traveling over the asphalt surface.

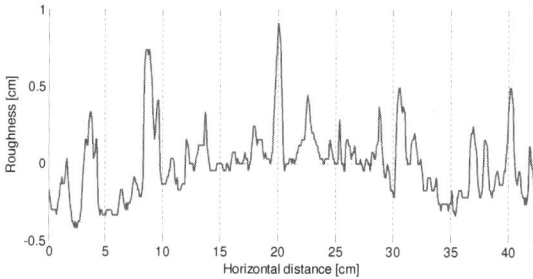

Figure 5.6: Asphalt profilometry by means of the high-precision optical laser sensor in a test carried out at constant vehicle speed v=3 km/h.

To better evaluate the influence of solar light, the asphalt profile was also measured by means of a precision laser triangulator run at very low speed (3 km/h) over the asphalt. This high precision optical sensor provides as output a voltage signal directly proportional to the asphalt roughness. The resulting roughness data, shown in Figure 5.6, were recorded and used to estimate the spectral distribution of the solar disturbances as a function of the vehicle speed.

Figure 5.7 shows the calculated relative noise spectra due to asphalt roughness at different vehicle speeds; namely 1 km/h, 15 km/h and 50 km/h. As can be seen, for speed values higher than 15 km/h, the noise contribution due to asphalt roughness exhibits significant spectral components at frequencies comparable to and higher than the LED modulation frequency of 1 kHz.

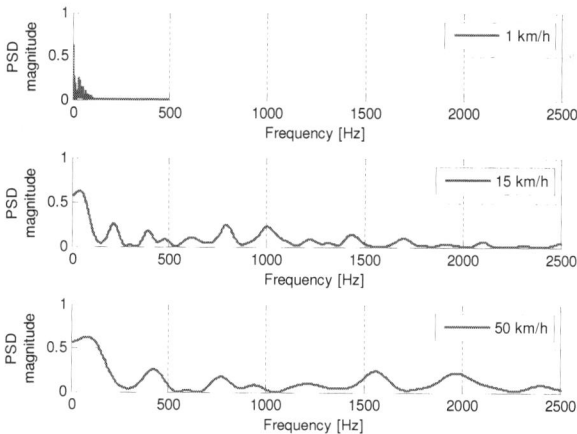

Figure 5.7: Relative power spectral density at different speeds.

A possible solution to this problem consists in achieving a better signal-to-noise ratio for the optical sensor. The adopted techniques for this purpose are described in the next Section.

5.3.1 Sensor design and characterization

To overcome the problems described in the previous Section, the low-cost commercial sensors were modified and customized to obtain better S/N conditions. In so doing, the main constraints are those of maintaining the small size and weight, and the low cost of the original sensor. Specifically, in order to improve the signal power, the LED emitter was replaced by a collimated laser diode in the near infrared. The new laser is still compact and low-cost, as the LED, but it provides a 10× boost in the optical power. The laser power supply is obtained from the LED supply by a simple current shunt; a parallel capacitor has also been added to protect the laser diode from current spikes. To significantly reduce the optical interference, the sensor was covered with a plastic optical filter (different from that used with the LED emitter): the chosen colored plastic foil exhibits a power attenuation of more than a factor 10 for the solar light spectrum, while optical attenuation is only 10 % in the near infrared (just the sum of the two facets reflection losses). These two improvements resulted in an optical signal-to-noise ratio increased of about 20 dB, without significantly modifying the whole instrument cost, size and weight.

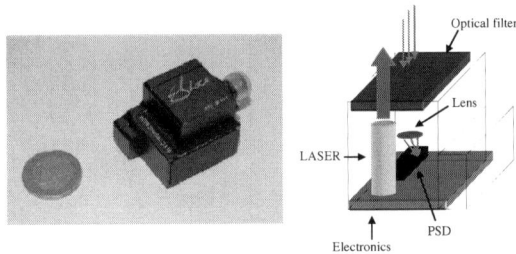

Figure 5.8: Photograph showing the final sensor and schematic description of the sensor internal structure.

The final sensor is mounted in a custom-made aluminum case, designed according to the IP67 requirements and so capable of withstanding the high vibrations of the motorcycle. The metal case is also helpful to shield the measured signal against external electrical noise. Shielded cables are used to transmit the detected signal out of the sensor, so as to avoid that the measurement is corrupted from radiated electromagnetic coupling which may occur in the "noisy" motorcycle environment. A photograph and a schematic description of the final sensor are shown in Figure 5.8.

The optical sensors developed were calibrated in the Lab using as a reference a commercial high precision laser triangulator (Baumer, Mod. OADM 13I6575/S35A), with resolution ranging from 10 µm to 0.4 mm in its distance measurement range from 10 cm to 100 cm. By means of quasi-static tests, performed moving a target, kept at constant speed, towards the sensor it was possible to experimentally measure the sensor curve, *i.e.*, the mapping between digital output voltage and measured distance (Figure 5.9). This curve will be useful in the uncertainty analysis to perform the optimization of the mounting parameters.

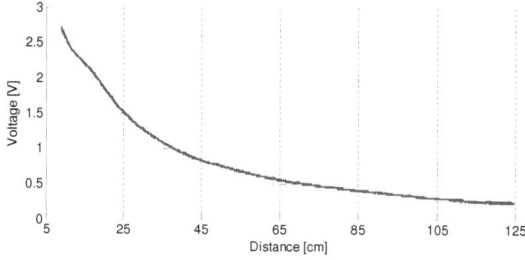

Figure 5.9: Distance-Voltage map of the designed triangulation telemeter.

5.3.2 Frequency response analysis

For the analysis of the frequency response of the used triangulation telemeter, it has to be considered that the measurement output has a refresh frequency of 25 Hz and the output sample is the mean of the data acquired in 40 ms: this is a method for filtering the high frequency noise components [68]. The low level processing of the PSD signal (the position x of the received beam on the PSD) is going to be explained to describe the frequency response of the electro-optical device.

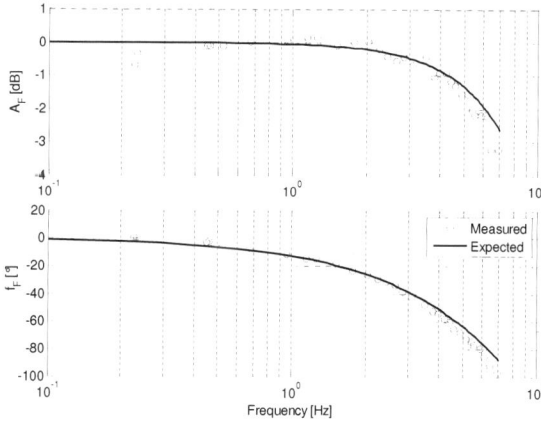

Figure 5.10: Frequency response of the triangulation telemeter for measuring the road roll angle of a motorcycle.

The PSD signal is sampled with a frequency of 1 kHz equal to the amplitude modulation, then it is integrated for a period of T_I=40 ms (*integration interval*) and the mean is computed. It can be shown that the application of this filtering method is equivalent to a filter which amplitude A_F and phase f_F are defined by

$$A_F(\omega) = \frac{\sin\left(\omega\frac{T_I}{2}\right)}{\omega\frac{T_I}{2}} ,$$

$$f_F(\omega) = -\omega\frac{T_I}{2}$$

(5.5)

where ω is the angular frequency.

In Figure 5.10 the frequency response of the triangulation telemeter is depicted: the black line represents the frequency response defined by Equation (5.5) and the blue circle represents the frequency response measured with a crank-rod system (Figure 5.11). The differences between the measured frequency response and the expected one are mainly due to the friction of the mechanical system.

Figure 5.11: Crank-rod system to analyze the frequency response of the presented triangulation telemeter for motorcycle roll angle measurement.

The electro-optical sensor has an amplitude attenuation of -3 dB at 7 Hz and a phase shift of -45° around 3 Hz. These performances are not critical if the electro-optical measurement system is used to measure the motorcycle lean angle.

5.3.3 Signal conditioning and processing

Once the two optical sensors are installed on the motorbike, at each time instant the measurements $V_1(t)$ and $V_2(t)$ are available. Through the best regression line of the characteristic curve of the sensor (see Figure 5.14, such measurements are converted into distances $d_1(t)$ and $d_2(t)$ and the road lean angle of the motorcycle can be estimated as in Equation (5.3). An example of the estimated tilt angle obtained via this expression is shown in the top plot of Figure 5.13.

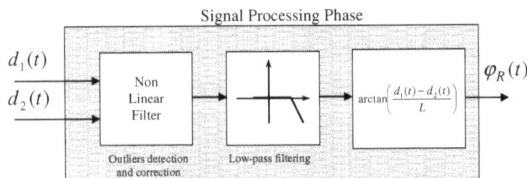

Figure 5.12: High level view of the signal processing strategy.

As can be seen considering the time intervals [64,66] s [70,72] s, some artefacts can still affect the tilt angle estimation (for example asphalt holes, rocks, manholes), which result in measurements outliers not related with the true vehicle tilt dynamic variations. To obtain a reliable and accurate measurement of the tilt angle, appropriate signal processing is needed. The high-level architectural view of this algorithm is shown in Figure 5.12.

The first step is to design a nonlinear filter to handle the outliers. Such a filter is comprised of an outlier detector and of a correction policy. To this purpose, at each time instant, $d_1(t)$ and $d_2(t)$ are checked as follows

$$\left| d_i(t) - d_i(t-1) \right| > h\Delta t, \; i = 1,2 , \tag{5.6}$$

where the value of the threshold h has been defined according to the maximum admissible rate of variation of the signal compatible with true tilt angle dynamics (a typical value is 0.5 cm/s), and Δt is the sampling time.

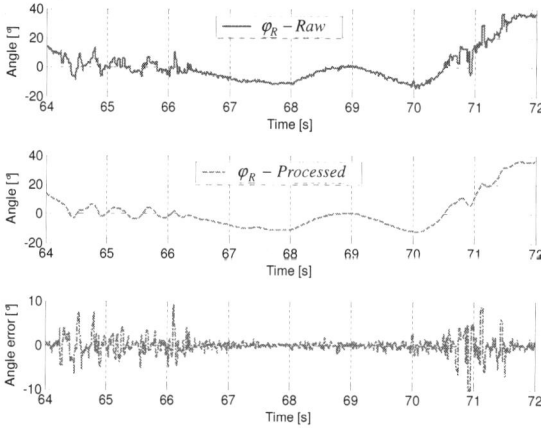

Figure 5.13: Plot of the raw tilt angle measurement (top), of the processed tilt angle (middle) and of the difference between the two signals (bottom), measured on a test track.

If one of the two distances fails to pass this consistency test its value $d_i(t)$ is discarded, and substituted with

$$\tilde{d}_i(t) = d_i(t-1) + sign\!\left(d_i(t) - d_i(t-1)\right)h\Delta t, \; i = 1,2 . \tag{5.7}$$

Finally, to reduce the effects of measurement noise, the two distance measures (after the outliers removal) are low pass filtered via a second order Butterworth filter with a cut-off frequency of 15 Hz. The filter bandwidth was designed considering that the tilt dynamics are in the range [0-

2] Hz and that the sensor bandwidth is about 7 Hz (see Section 5.3.2). The latter has been evaluated pointing the sensor towards a target moved with multi-tone sinusoidal motion and experimentally computing its frequency response, [69]. The final estimation of the tilt angle is shown in the middle plot of Figure 5.13. To better appreciate the importance of the signal processing phase, the bottom plot in Figure 5.13 shows the difference between raw and estimated tilt angle.

5.4 Uncertainty analysis

The uncertainty in the distance measure is mainly due to the quantization of the internal analog to digital converter (ADC) of the sensor. The microcontroller uses an ADC with 8 bits with a range [0-5] V, resulting in a quantization level of 20 mV.

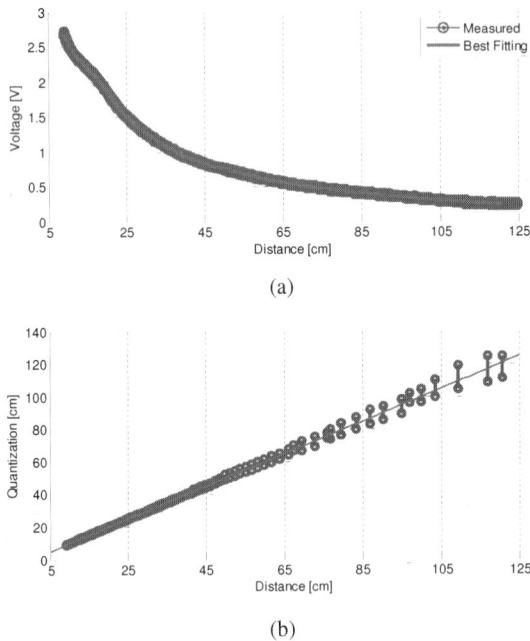

(a)

(b)

Figure 5.14: (a) Sensor characteristic curve (circles) and best polynomial fitting (solid line). (b) Calibration error as function of distance.

Figure 5.14 shows the sensor characteristic curve, mentioned in the previous section, together with its best fifth order polynomial fitting, obtained via least square regression. Analyzing the calibration error, *i.e.*, the difference between experimental data and the fitting curve, as a function of the measured distance - bottom plot of Figure 5.14 - shows the effect of the voltage quantization. One has to recall that the sensor output is inversely proportional to the measured distance.

The remaining part of this Section will be devoted to analyze how the voltage quantization error translates into a tilt angle measurement uncertainty. To this end, considering that the measured voltage uncertainty $u(V)$ is dominated by the quantization error, it is useful to express this quantity as

$$u^2(V) = \frac{LSB^2(V)}{12},$$ (5.8)

where $LSB(V)$ represents the value of the least significant bit of the measured voltage. This quantity is also the maximum error in the voltage measure.

If we employ the polynomial relationship between the measured distance and the measured voltage, shown in the top plot of Figure 5.14, the distance uncertainty $u(d)$ can be expressed as in [70]:

$$u^2(d) = \left(\frac{\partial d}{\partial V}\right)^2 u^2(V) = \left(\sum_{i=1}^{n} ic_i V^{i-1}\right)^2 u^2(V),$$ (5.9)

where $d = \sum_{i=0}^{n} c_i V^i$ is the polynomial expression of the distance as a function of voltage. From (5.8) and (5.9), the maximum distance error has the form

$$E_{MAX}(d) = \left(\sum_{i=1}^{n} ic_i V^{i-1}\right) LSB(V).$$ (5.10)

Employing Equation (5.3), the uncertainty and maximum error expressions as functions of the tilt angle φ_R and of the distance uncertainty $u^2(d_i)$, $i = 1, 2$ can be expressed as (see [70])

$$u^2(\varphi_R) = \sum_{i=1}^{2} \left(\frac{\partial \varphi_R}{\partial d_i}\right)^2 u^2(d_i) = \frac{\sum_{i=1}^{2} u^2(d_i)}{\left(L[1 + \tan^2(\varphi_R)]\right)^2}, \quad i = 1, 2$$

$$E_{MAX}(\varphi_R) = \sum_{i=1}^{2} \left|\frac{\partial \varphi_R}{\partial d_i}\right| E_{MAX}(d_i) = \frac{\sum_{i=1}^{2} E_{MAX}(d_i)}{L[1 + \tan^2(\varphi_R)]}, \quad i = 1, 2$$ (5.11)

The uncertainty expressions have been obtained under the hypothesis of uncorrelated measurement of d_1 and d_2, while the maximum error is calculated directly summing all the error contributions without signs.

Inspecting (5.11), it is possible to note that the both the tilt angle uncertainty and the maximum error depend on the mounting parameter L. As such, the mounting distance L between the two sensors can be adequately chosen to minimize the roll angle uncertainty $u^2(\varphi)$ and the maximum error $E_{MAX}(\varphi)$. This can be accomplished selecting the optimal value L^* as the minimum of the following cost function $J(L)$

$$L^* = \arg\min_L J(L) \geq \arg\min_L \{u(\varphi_R) + E_{MAX}(\varphi_R)\}. \tag{5.12}$$

Note, however, that the minimization of (5.12) must be carried out taking into account the following constraints on the value of the second mounting parameter H_0 (see Figure 5.3)

$$H_0 \text{ such that } \begin{cases} H_0 + \dfrac{L}{2}\tan(\varphi_{MAX}) \leq d_{MAX} \\ H_0 - \dfrac{L}{2}\tan(\varphi_{MAX}) \geq d_{min} \end{cases} \tag{5.13}$$

where d_{min} and d_{MAX} are the upper and lower limits of the sensors range, respectively, while φ_{MAX} is the maximum measurable value of the road roll angle. For the considered sensor, d_{min}=10 cm and d_{MAX}=100 cm. The value of φ_{MAX} was set to φ_{MAX}= 55°, which is consistent with the instrument application.

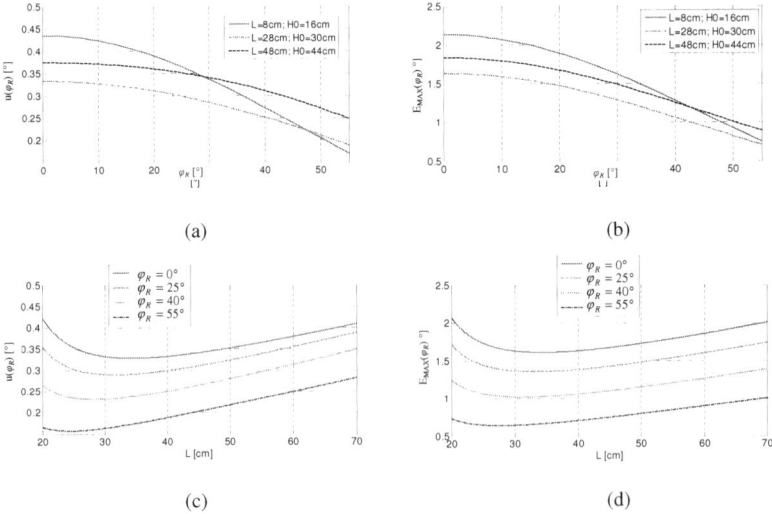

(a)

(b)

(c)

(d)

Figure 5.15: Plot of $u(\varphi_R)$ (a) and $E_{MAX}(\varphi_R)$ (b) for different choices of L and H_0 and with φ_{MAX}= 55° as functions of the tilt angle; plot of $u(\varphi_R)$ (c) and $E_{MAX}(\varphi_R)$ (d) as functions of the mounting distance L for different values of the tilt angle and with H_0 computed via (5.14).

In order to optimally exploit the sensor characteristic, it is desirable to work in the range that corresponds to the portion of the characteristic with the higher sensitivity. Hence, the optimal value of H_0 is computed based on the second constraint in (5.14) as

$$H_0^* = d_{min} + \frac{L^*}{2}\tan(\varphi_{MAX}) \tag{5.14}$$

so as to guarantee that the distance range in which the sensors are used is spanned starting from the minimum available distance from ground.

Thus, the following nonlinear constrained optimization problem

$$L^* = \arg\min_L J(L) = \arg\min_L \{u(\varphi_R) + E_{MAX}(\varphi_R)\}$$

such that

$$H_0 + \frac{L}{2}\tan(\varphi_{MAX}) \le d_{MAX}$$

$$H_0 = d_{min} + \frac{L}{2}\tan(\varphi_{MAX})$$

(5.15)

has been solved to give L^* and H_0^*.

Figure 5.15a and Figure 5.15b show, respectively, the behavior of the functions $u(\varphi_R)$ and $E_{MAX}(\varphi_R)$ for different choices of L and H_0 computed for $\varphi_{MAX}=55°$, while Figure 5.15c and Figure 5.15d show again $u(\varphi_R)$ and $E_{MAX}(\varphi_R)$, plotted as functions of the mounting distance L, for different values of the tilt angle and with H_0 computed via (5.14) for each value of L. These figures show that, for the considered sensors and the fixed value of maximum tilt angle, the optimal mounting parameter values would be $L^*=28$ cm and $H_0^*=30$ cm.

A major difficulty in the experimental set-up is caused by the lack of available mounting positions for the sensors. The best final position, determined by trading off constraints due to measurement resolution optimization with the available mounting points, was found to be either the hooking of the rear stand of the motorbike or below the pedals, as previously shown in Figure 5.4.

5.5 Pitch and roll angle relation

Besides the measurement uncertainty due to quantization errors, in the considered application one needs to take into account also dynamic uncertainties due to the vehicle non planar motion, [28], [4], [7]. One of the most significant distortions in the tilt angle measurement is given by the effect of the pitch angle dynamic, caused by acceleration, braking, and weight distribution. Hereafter, ϑ_R will denote the pitch angle of the motorcycle measured with respect to the road (*road pitch angle*).

To evaluate the effects of the pitch disturbance on the tilt angle measure, Equation (5.4) can be rewritten considering also a non-zero pitch angle.

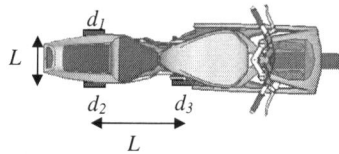

Figure 5.16: Motorcycle set-up for roll and pitch measurement.

Consider the motorcycle equipped with three triangulation telemeters as depicted in Figure 5.16. The attitude angles φ_R and ϑ_R are yield by

$$
\begin{aligned}
\vartheta_R &= \arctan\left(\frac{d_2 - d_3}{L_\vartheta}\right) \\
\varphi_R &= \arctan\left(\frac{(d_1 - d_2)\cos(\vartheta_R)}{L_\varphi}\right)
\end{aligned}
\tag{5.16}
$$

where L_φ is the distance between the rear sensors and L_ϑ is the distance between the sensors mounted on the motorcycle side. The measured distances can be modeled as

$$
\begin{cases}
d_1 = \left[H_0 + \dfrac{L}{2}\tan(\varphi_R)\right]\dfrac{1}{\cos(\vartheta_R)} + B(\vartheta_R)\tan(\vartheta_R) \\[2mm]
d_2 = \left[H_0 - \dfrac{L}{2}\tan(\varphi_R)\right]\dfrac{1}{\cos(\vartheta_R)} + B(\vartheta_R)\tan(\vartheta_R) \\[2mm]
d_3 = \left[H_0 - \dfrac{L}{2}\tan(\varphi_R)\right]\dfrac{1}{\cos(\vartheta_R)} + (B(\vartheta_R) - L_\vartheta)\tan(\vartheta_R)
\end{cases}
\tag{5.17}
$$

where $B(\vartheta_R)$ is the distance between the instantaneous center of rotation of the pitch dynamics and the optical sensors.

If the pitch angle is not compensated as in (5.16), a measuament error is committed. Considering a roll angle range of 50° and a pitch range of 15° the measurement error is less than 1° (Figure 5.17).

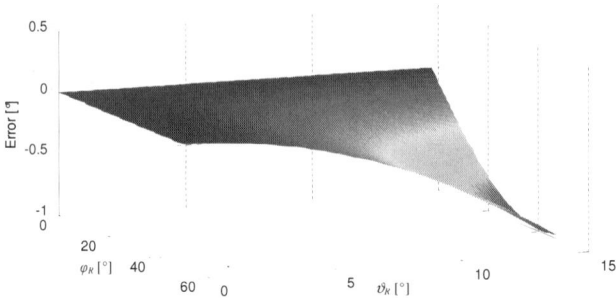

Figure 5.17: Error introduced by the pitch dynamic in the roll angle measurement with electro-optical sensors.

Considering Equation (5.16) and Equation (5.17), the roll angle measurement uncertainty is given by Equation (5.18).

Considering the best values of the mounting parameters given above, the roll uncertainty given in (5.20) can be computed with the expression of the distances d_i, $i = 1, 2$ in (5.17) to provide an indication of the pitch effects on the tilt angle measure quality. Figure 5.18 shows a 3D plot of $u(\varphi)$ as a function of both pitch and tilt angles, calculated for $B(\vartheta_R) = 60$ cm.

$$u^2(\vartheta_R) = \frac{\sum_{i=1}^{2} u^2(d_i)}{\left(L_\vartheta\left[1 + \tan^2(\vartheta_R)\right]\right)^2}, \ i = 2, 3$$

$$\Phi = \frac{(d_1 - d_2)\cos(\vartheta_R)}{L_\varphi}$$ (5.18)

$$u^2(\Phi) = \frac{\cos^2(\vartheta_R)}{L_\varphi^2} \sum_{i=1}^{2} u^2(d_i) + \Phi^2 \tan^2(\vartheta_R) u^2(\vartheta_R), \ i = 1, 2$$

$$u^2(\varphi_R) = \left(\frac{1}{1 + \Phi^2}\right)^2 u^2(\Phi)$$

Note that:

- The range of pitch angle variation considered in Figure 5.18 is not symmetrical. This is due to the fact that much larger pitch angles occur during braking (where the pitch angle is positive) rather than in acceleration (where the pitch angle is negative)
- The considered attitude angle are referred to the road (φ_R and ϑ_R), because these are the angle that are measurable by the electro-optical system.

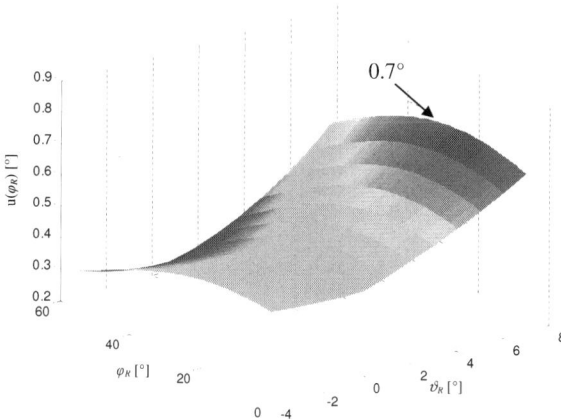

Figure 5.18: Plot of the tilt angle measurement uncertainty $u(\varphi)$ as a function of pitch and roll angles.

Figure 5.18 also shows that the overall tilt angle measurement uncertainty is always below 0.7°, which can be regarded as a very good performance for the considered application. The accuracy of the electro-optical system can be further improved considering a redundant sensors configuration.

5.6 Analysis of tire thickness

The aim of this Section is to show that the tires thickness does not influence the electro-optical measurement of the motorcycle lean angle.

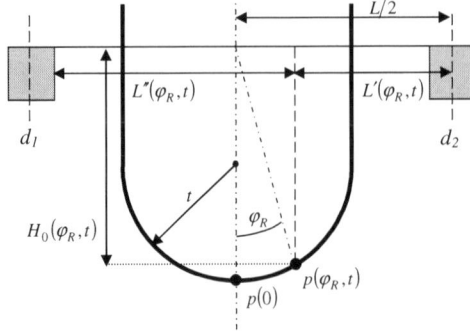

Figure 5.19: Representation of the relation between the point of contact of a tire and the road roll angle of a motorcycle.

For the sake of conciseness, consider the road roll angle measurement if the pitch angle dynamic can be neglected. In Figure 5.19, it is depicted how the point of contact (POC) of the tire $p(\varphi_R, t)$ changes due to the tire thickness t and the road roll angle φ_R. If the tire thickness is null, then, the POC lies on the symmetrical axis of the motorbike and the distance between the axis passing through the point of contact and the sensors (named $L'(\varphi_R, t)$ and $L''(\varphi_R, t)$) does not depend on φ_R and are always equal to $L/2$.

Otherwise if $t > 0$, $L'(\varphi_R, t)$ and $L''(\varphi_R, t)$ are defined by

$$L'(\varphi_R, t) = \frac{L}{2} - t\sin(\varphi_R)$$
$$L''(\varphi_R, t) = L - L'(\varphi_R, t) = \frac{L}{2} + t\sin(\varphi_R)$$

$$(5.19)$$

From Figure 5.3 and Figure 5.19, it can be easily deduced that

$$\begin{cases} d_1 = H_0(\varphi_R, t) + L''(\varphi_R, t)\tan(\varphi_R) \\ d_2 = H_0(\varphi_R, t) - L'(\varphi_R, t)\tan(\varphi_R) \end{cases}$$

$$(5.20)$$

and

$$d_1 - d_2 = L\tan(\varphi_R),$$

$$(5.21)$$

thus, the road roll angle measured by the electro-optical measurement is not affected by the tire thickness.

5.7 Experimental test

The final experimental tests on an instrumented motorbike were carried out on a hyper-sport class motorcycle (Aprilia TUONO1000). The motorcycle is equipped with an on board Electronic Control Unit (ECU) with both analog and digital inputs sampled up to 1 kHz. The analog inputs can be acquired at 10 bit or 12 bit resolution in a range 0-5 V. In our setting, the sensors were connected to the 10 bit input channels (5 mV resolution) because the internal quantization level of the sensor does not allow to exploit the 12-bit resolution of the acquisition board. Two sensors were placed under the pedals and the final mounting parameter values on board of the vehicle were $L= 40$ cm and $H_0= 34$ cm.

The algorithm described in the previous Sections has been implemented on board of the motorcycle ECU, where the polynomial fit of the static sensor calibration curve was also stored. This implementation allows for a real-time roll-angle computation from the measurement outputs of the two sensors. The first test consisted of a high-speed trial (> 300 km/h) on a straight track segment, in a very sunny day. Under such extreme conditions, the sensors provided consistent measurements over the complete acquisition interval, demonstrating the immunity of the new optical design to the problem of solar interference. Finally, the roll-angle measurement system was tested on a race-track (Misano circuit). Figure 5.20 shows the map of the race-track and the roll angle measurements recorded during a whole lap. The points corresponding to track curves are reported in both figures, to clearly describe the motorcycle attitude during the lap. As expected from previous tests, the novel measurement system exhibits very good performances also for a racing application, with adequate dynamics, accuracy, and reliability.

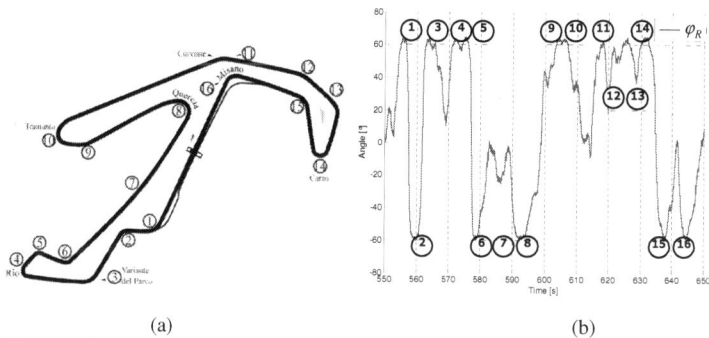

(a)   (b)

Figure 5.20: Map of the Misano circuit (a) and tilt angle measurement over one lap (b). The curves in the circuit are numbered and correspondingly indicated in the tilt angle time history.

5.8 Concluding remarks

A novel optical sensor for tilt angle estimation in hypersport motorcycles was studied, designed, developed and tested. From the characterization work it is clearly demonstrated that the measurement system using a commercial triangulator based on LED emitters suffers too much from solar interference and it is not adequate for real-time measurements at high motorbike speeds. To overcome the reliability problems due to solar interference, a novel sensor using a laser source emitter (more powerful and with narrower optical band, as compared to the LED emitter) and an optical filter was designed. The first measurements with the new laser sensor showed excellent performance on the track test at different speeds (0-300 km/h) and even under strong solar lightning conditions. A set of the new optical sensors are currently being tested, mounted on racing motorbikes, and undergoing pre-competition tests performed in collaboration with leading racing teams.

Chapter 6
Roll angle estimation via position signals

In this Chapter the estimation of the roll angle of a motorbike with position signals is presented. Until few years ago, the GPS estimation of the lean angle has been the only way to obtain a representative measurement of the motorcycle inclination. Consequently, this Chapter is devoted to the explanation of the estimation principle and the main drawbacks of the method.

First of all, the GPS system will be presented (Section 6.1), then, the experimental set-up is described (Section 6.2) and the problem of roll angle estimation will be tackled (Section 6.4). The estimation of the lean angle of a motorbike with position signals requires the estimation of the radius curve of the covered track. As a consequence in Section 6.3 the problem of curve radius estimation will be taken into account.

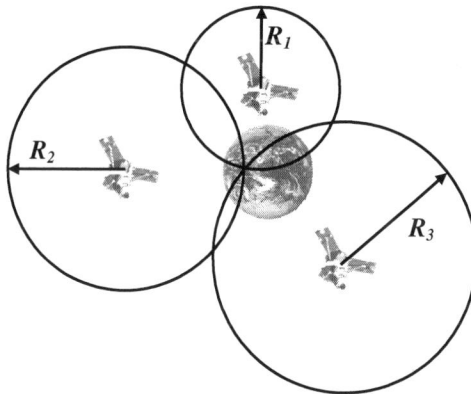

Figure 6.1: Basic idea of GPS positioning.

6.1 GPS system

Herein a brief description of how a GPS works and of which are the principle terms of error is proposed.

The *Global Positioning System* (GPS) is a satellite based navigation system firstly developed for military system and it was later available for civilian usage. The standard is well described in text books such as [71], [72], [73], [74].

The GPS system is adopted both to measure the absolute position and the absolute velocity of a receiver on the Earth surface. The position is measured as *pseudorange observations*, while the velocity is measured though *Doppler measurements*.

Pseudorange observations are obtained by measuring the how long it takes for the signal to propagate form the satellite to the receiver and multiplying the propagation time by the speed of light (c).

If pseudorange measurements can be made from at least four satellites, enough information exists to solve the positioning problem with GPS signals. In particular, three satellites are necessary to define the position of the receiver on the earth surface using trilateration (Figure 6.1) and one satellite is needed to solve the time synchronization problem [12].

The pseudorange measurement for a single satellite can be expressed as:

$$\rho = P + dP + c\left(dt - dT\right) + d_{ion} + d_{trop} + \varepsilon_\rho \tag{6.1}$$

where:

- ρ is the pseudorange observation
- P is the true range between the GPS satellite and the receiver
- dP is the orbital error of the satellite
- dt is the satellite clock error
- dT is the receiver clock error
- d_{ion} is the delay due to the ionosphere
- d_{trop} is the delay due to the troposphere
- ε_ρ is the measurement noise and multipath effect

The main source of error is the receiver clock error that causes a biased range measurement. The effect of this error can be reduced considering four satellites to resolve the positioning problem instead of just three.

Doppler measurements are obtained by determining the change in phase over a given time interval dived by the interval length [12]. A change in phase corresponds to a frequency shift that is proportional to the relative velocity between the emitter and the receiver: this is commonly known as Doppler effect. As a consequence, knowing the emitter velocity and measuring the frequency shift, the absolute receiver velocity can be determined. As pseudorange measurements, also Doppler measurement are corrupted by several sources of error:

$$\dot{\rho} = \dot{P} + d\dot{P} + c\left(d\dot{i} - d\dot{T}\right) + \dot{d}_{ion} + \dot{d}_{trop} + \dot{\varepsilon}_{\rho} \tag{6.2}$$

where:

- $\dot{\rho}$ is the Doppler observation
- $\dot{P}$ is the true range rate between the GPS satellite and the receiver
- $d\dot{P}$ is the orbital error drift of the satellite
- $d\dot{i}$ is the satellite clock drift
- $d\dot{T}$ is the receiver clock drift
- $\dot{d}_{ion}$ is the delay drift due to the ionosphere
- $\dot{d}_{trop}$ is the delay drift due to the troposphere
- $\dot{\varepsilon}_{\rho}$ is the measurement noise and multipath effect.

In order to evaluate the combined effect of the errors described above, each one of them is converted in the so called *User Equivalent Range Error* (UERE). In general, errors from different sources have different statically properties. For example the errors due to satellite clock and ephemeris tend to vary slowly and tend to appear as biases over long time interval. On the contrary, errors due to receiver noise and quantization effect may vary much more rapidly. However, if sufficiently long time is considered, all errors can be assumed as independent zero-mean random process that can be combined to form a single UERE (Equation (6.3); see [75]).

$$UERE = \sqrt{\sum_{i=1}^{n} \left(UERE_i\right)^2} \ . \tag{6.3}$$

Segment	Error sources	*1-σ* UERE [m]
Space	Stability of satellite clock	3.0
	Satellite perturbation	1.0
	Others	0.5
Control	Ephemeris errors	4.2
	Others	0.9
User	Ionospheric delay	5.0
	Tropospheric delay	1.5
	Multipath	2.5
	Measurement noise	1.5
	Others	0.5
UERE	Total	8.0

Table 6.1: Typical GPS UERE budget [73].

In Equation (6.3), the *1-σ* UERE of all sources of errors (*UERE_i*) are combined as sum of the squared values: the combined UERE represents the measurement uncertainty of the quantity measured with the GPS signals. Typical UERE values are reported in Table 6.1 [73].

Other than the description of the error with the UERE model, other characterizations have been proposed. In [74], the elevation angle El_i of each satellite is taken into account and it is assumed that the measurement noise for satellite *i-th* is of the form

$$w_i(El_i) = w_{0,i} + w_1 f_i(El_i),$$
$$\hspace{8cm}(6.4)$$

where $w_{0,i}$ are independent noise terms of variance $7\sigma^2/8$, w_1 has variance $\sigma^2/8$ and the obliquity factor f_i is approximated as

$$f_i(El_i) = \frac{1.1}{\sin(El_i) + 0.1}, \; El_i > 0°.$$
$$\hspace{6cm}(6.5)$$

6.2 Experimental set-up

The performance of the estimation of the motorcycle lean angle with position signals has been tested with data collected on the Mugello circuit (Figure 4.9c). The vehicle has been instrumented with an high performance GPS receiver with an internal IMU and two electro-optical sensors. The data provided by the electro-optical instrumentation is adopted to measure the lean angle of the motorcycle with respect to the road as described in Chapter 5.

The GPS module has a refresh time of 0.1 ms; in Table 6.2 the accuracy of the employed device is reported.

Measurement	UERE
Absolute position	2.5 m
Velocity	0.6 m/s

Table 6.2: GPS Module accuracy.

6.3 Radius curve estimation

The first step of the lean angle estimation of the motorcycle with GPS signals is the estimation of the curvature radius in every point of the circuit. In [76] two methods are presented to compute the curvature radius of the circuit.

Consider a data set constituted by N pairs (x_i, y_i), $i=1,...,N$, which represents 2-D coordinates in the fixed reference frame (if the altitude coordinate is also available, then the procedures can be optimized). A first approach is based on the estimation of the circumference that better fits a vector of N position points (Figure 6.2a); a second approach consists on the estimation of the yaw angle variation between two consecutive position points (Figure 6.2b).

6.3.1 Circle fitting method

Consider a vector of N pairs of 2-D position measurements. The solution of the fitting problem can be referenced to [77] in which the solution of least-square fitting of a circle is proposed: the geometric error (sum of squared distances from the points to the fitted circle) is minimized using nonlinear least squares based on Gauss-Newton method. Once the parameters of the circle equation are obtained, then, the curvature radius R_i can be computed.

It is well known that at least three points are necessary to define a circle and if $N>3$, then, a spatial filter is applied to the estimation algorithm: the greater is the value of N, the smoother is the curvature estimation. For the roll angle estimation problem it is considered $N=10$.

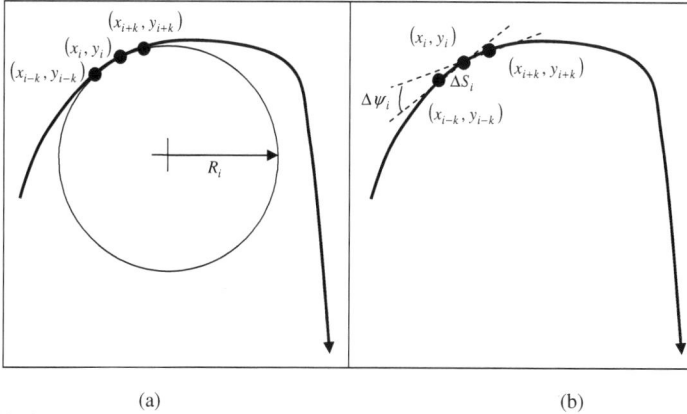

(a)   (b)

Figure 6.2: Curvature radius estimation. (a) Circumference fitting. (b) Yaw angle variation estimation.

6.3.2 Yaw angle variation method

Consider that $\Delta \psi_i$ is the angle between two adjacent points and Δs_i is the distance between them. Consequently, the estimated radius curvature r_i can be computed as

$$R_i = \left| \frac{\Delta \psi_i}{\Delta s_i} \right|, \tag{6.6}$$

where

$$\Delta \psi_i = \arctan\left(\frac{y_i}{x_i}\right) - \arctan\left(\frac{y_{i-1}}{x_{i-1}}\right). \tag{6.7}$$

$$\Delta s_i = \sqrt{\left(x_i^2 + y_i^2\right) - \left(x_{i-1}^2 + y_{i-1}^2\right)}$$

Both of the estimation methods provide a very noisy initial estimate, thus, to obtain a smooth curvature radius the initial estimate needs to be low-pass filtered. In Figure 6.3 the comparison of the estimation methods is depicted. The reference radius curvature is provided by the adopted GPS module.

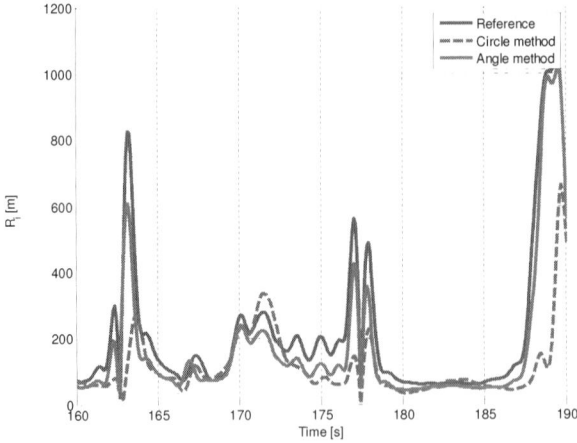

Figure 6.3: Comparison of the curvature radius estimation methods.

The curvature radius estimated with the circle fitting method has worse performance. This is due to the noise that afflicts the acquired position signals and to the fact that this approach is less robust than the method based on the yaw angle variation. As a consequence, the estimation of the lean angle of the motorcycle will be based on the curvature radius estimated through Equation (6.6) and (6.7).

6.4 Roll angle estimation

Once the curvature radius of the motorcycle trajectory has been estimated, the lean angle of the vehicle can be estimated applying Equation (4.4) as

$$\varphi = \arctan\left(\frac{V_x^2}{Rg}\right) \tag{6.8}$$

where V_x is the longitudinal speed and R is the curvature radius. The main drawbacks of this expression have been already underlined in Chapter 4.

$$\hat{\varphi}_{GPS} = \arctan\left(\frac{V_x^2}{R_i g}\right)$$
$$R_i = \left|\frac{\Delta \psi_i}{\Delta s_i}\right| \tag{6.9}$$

Combining Equations (6.6) and (6.8), the roll angle can be estimated through position signals as in Equation (6.9).

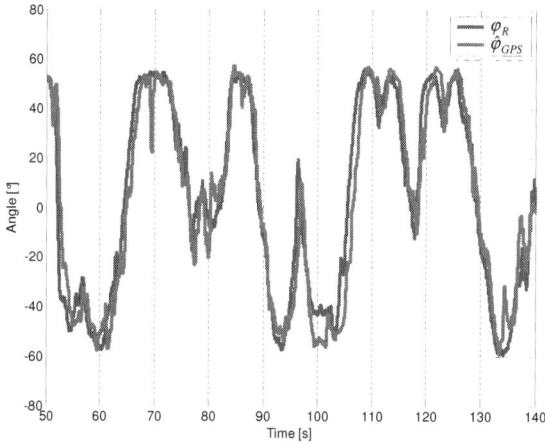

Figure 6.4: Roll angle estimation with position signals.

In Figure 6.4 a comparison between the reference lean angle obtained by the electro-optical measurements and the roll angle estimated by (6.9) is depicted. The estimation performance is defined by the ESR (Error to Signal Ratio) that is the ratio between the MSE (Mean Square Error) of the estimation error and the MSE of the reference signal in one lap. The ESR of the roll angle estimation with GPS signals in the Rijeka circuit is about 15%. Even if the estimation performance is not critical, the main problem of the estimation of the lean angle with position signals is the delay due to GPS that is evident in the between 100 s and 110 s.

6.5 Concluding remarks

In this Chapter the estimation of the lean angle of a motorcycle with position measurements has been presented. This is interesting because it represent the state of the art in many racing applications. It has been shown that it is based on a steady state expression of the lean angle (Equation (6.8)), thus, the estimation performance are strongly influenced by the limits of this formula that can be overcome limiting the estimation of the lean angle to the low frequency component of the attitude parameter. Often an indirect Kalman filtering formulation based on inertial and position signals is adopted to reduce the ESR of the estimation and avoid the effect of delay of the GPS system on the estimated quantities.

In the following Chapters the estimation of the roll angle with inertial sensors will be presented.

Chapter 7
Motorcycle attitude estimation with inertial sensors: problem setting

In this Chapter the estimation of the attitude angles of a motorcycle via inertial sensors is introduced and some basic concepts that are useful for designing estimation algorithms are given. The problem of the estimation of a the attitude of a motorcycle with inertial signals is not trivial mainly because of the very noisy environment in which the data are collected and the errors that affects the signals provided by the MEMS sensors.

As stated in Chapter 1, three chapters are devoted to this problem. Herein the general solution of the problem is going to be introduced detailing the main limits of the inertial sensors and how the acquired signals need to be treated before they are used as inputs to the attitude estimation algorithms.

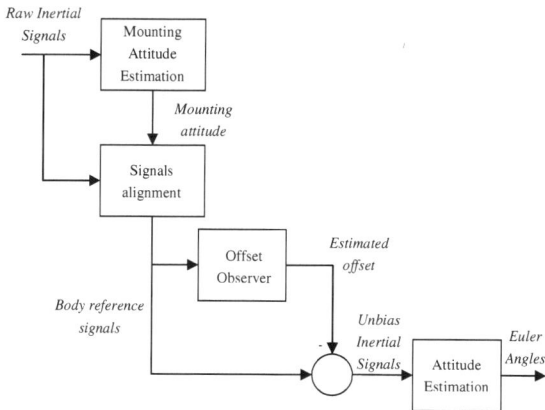

Figure 7.1: High-level estimation scheme with offset observer.

The problems that need to be solved are:

- Model of the acquired inertial signals
- Alignment of the inertial platform with the body coordinate system
- Offset compensation
- Attitude estimation.

In this work, the complexity of the problem is split into different simpler blocks: the high level scheme in Figure 7.1 is proposed.

The problem of signals alignment, offset estimation and attitude estimation are treated separately by three different algorithms, consequently the input inertial signals to the attitude estimation algorithm are considered to be aligned in the body reference frame and unbiased.

The Chapter is organized as follows. First of all, the models of the acquired signals are introduced: these are fundamentals to design estimation algorithms (Section 7.1). At the end of the Chapter, the equations of an inclinometer to measure the attitude of the vehicle in a static condition and the offset estimation algorithm are introduced (Sections 7.2 and 7.3).

7.1 Model of the acquired signals

In this Section the models of the inertial signals that are acquired will be presented. Both the models of the measured angular rates and accelerations will be based on a kinematic expression of the signals adding a description of the errors that affect the measurements.

The models of the inertial signals are fundamentals for developing algorithm for attitude estimation. All the proposed algorithms are model based, and then the performance strictly depends on the hypothesis of the adopted model.

As stated in Chapter 3, the definition of the angular rates and the expression of the accelerations depends on the description of the orientation of the body. In this application Euler angles are used and in particular the set **Z-X-Y** is adopted (to the aim of conciseness, in this Section the subscript E will be omitted). It means that at each instant the orientation of the body is defined as a sequence of rotation:

- *First rotation*: angle ψ along the *yaw* axis;
- *Second rotation*: angle φ along the *roll* axis;
- *Third rotation*: angle ϑ along the *pitch* axis.

Therefore, the roll angle is defined as the rotation along the X axis that is necessary to realign the relative and absolute Z axis when the pitch angle is null. As a consequence, to improve the performance of the lean angle estimation, it is also necessary to estimate and compensate the pitch angle of the motorbike. This fact will be remarked in the following Chapters dedicated to the design of estimation algorithms.

The coordinate reference system that are employed to obtain the interesting models are: the *inertial reference frame* (**XYZ**) and *Body reference frame* (**xyz**) already defined in Chapter 4 and the *intermediate reference frame* (**XYZ**)*'* that is a dextral time variant coordinate system positioned in the projection of the center of gravity (COG) on the line corresponding to the projection of the symmetry plane of the vehicle on the road plane and oriented as the inertial reference frame.

In the following it is assumed that the inertial platform is positioned in accordance with the body frame, moreover without loss of generality, it will be considered that the IMU is mounted in the

COG of the vehicle. The kinematics quantities described in the inertial reference frame will be indicated with capital letters, while the kinematics quantities described in the body reference frame will be indicated with minor letters. In Figure 7.2 the above coordinate systems are depicted.

The rotation between *(XYZ)* and *(xyz)* defined by the chosen Euler angles is described by the rotation matrix (see Chapter 3)

$$R_{ZXY}(\varphi, \vartheta, \psi) = R_Y(\vartheta)R_X(\varphi)R_Z(\psi) =$$

$$= \begin{bmatrix} c_\vartheta c_\psi - s_\varphi s_\vartheta s_\psi & c_\vartheta s_\psi + s_\varphi s_\vartheta c_\psi & -c_\varphi s_\vartheta \\ -c_\varphi s_\psi & c_\varphi c_\psi & s_\varphi \\ s_\vartheta c_\psi + s_\varphi c_\vartheta s_\psi & s_\vartheta s_\psi - s_\varphi c_\vartheta c_\psi & c_\varphi c_\vartheta \end{bmatrix} \qquad (7.1)$$

where $R_x(\varphi)$, $R_y(\vartheta)$ and $R_z(\psi)$ represent respectively the elementary rotation around the X axis of an angle φ, the Y axis of an angle ϑ and the Z axis of an angle ψ, defined in Chapter 3.

The adopted representation of the orientation of the vehicle avoids that in the kinematic problem the singularities of the rotational matrix $R_{ZXY}(\varphi, \vartheta, \psi)$ belong to the range of variation of the attitude angles [27] for the motorcycle application, consequently, the Euler angle representation does not introduce a range limit with respect to quaternion based representation.

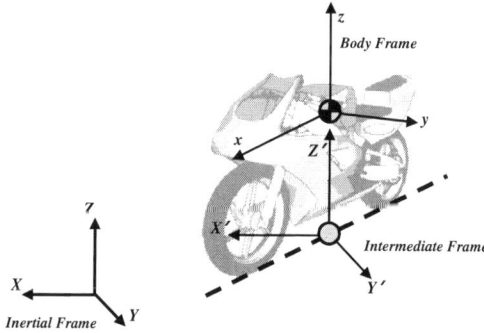

Figure 7.2: Reference frame definition and orientation of the axes.

7.1.1 Model of the measurements of the gyroscopes

The signals provided by the MEMS angular rate sensors are composed by two terms:

1. Kinematic contribution
2. Noise contribution.

To obtain the expression of the kinematic angular rates of the vehicle, the kinematic approach described in Chapter 3 and also reported in [3], [78] and [79] is applied, thus, for the adopted set of the Euler angles, Equation (7.2) is achieved (the measured angular rates are represented by ω_x, ω_y and ω_z).

$$\begin{bmatrix} \omega_x \\ \omega_y \\ \omega_z \end{bmatrix} = \begin{bmatrix} c_\vartheta \dot{\varphi} - s_\vartheta c_\varphi \dot{\psi} \\ \dot{\vartheta} + s_\varphi \dot{\psi} \\ s_\vartheta \dot{\varphi} + c_\varphi c_\vartheta \dot{\psi} \end{bmatrix} \qquad (7.2)$$

A widely used model to describe the noise of the signal acquired by a MEMS gyroscope is

$$\tilde{\omega}(t) = \omega(t) + \Delta_\omega(t) + \eta_\omega(t)$$
$$\dot{\Delta}_\omega(t) = \eta_{\Delta_\omega}(t) \qquad (7.3)$$

where $\tilde{\omega}(t)$ is the continuous-time measured angular rate, $\Delta_\omega(t)$ is the model of the offset of the sensor and $\eta_\omega(t)$ and $\eta_{\Delta_\omega}(t)$ are independent zero mean Gaussian white noise process with

$$E\left[\eta_\omega(t)\eta_\omega^T(\tau)\right] = \sigma_\omega^2 \delta(t-\tau)$$
$$E\left[\eta_{\Delta_\omega}(t)\eta_{\Delta_\omega}^T(\tau)\right] = \sigma_{\Delta_\omega}^2 \delta(t-\tau) \qquad (7.4)$$

where $\delta(t-\tau)$ is the Dirac delta function.

7.1.2 Model of the measurements of the accelerometers

The signals provided by MEMS accelerometers, is often used just to represent the direction of the gravitational acceleration with respect to the attitude of the vehicle [80], [81] or a network of accelerometers can be used to replace gyros [82], [83]. In a motorcycle application this simplification is not possible, thus, the accelerations of the vehicle that are measured by a MEMS sensor are considered to be composed by:

1. Kinematic acceleration
2. Gravitational acceleration
3. Noise.

The kinematic accelerations a_x, a_y and a_z of the COG of the vehicle can be expressed as

$$\begin{bmatrix} a_x \\ a_y \\ a_z \end{bmatrix} = R_{ZXY}(\varphi, \vartheta, \psi) \begin{bmatrix} A_X \\ A_Y \\ A_Z \end{bmatrix} \qquad (7.5)$$

where A_X, A_Y and A_Z are the component of the kinematic acceleration vector along the X, Y and Z inertial reference axis

Referring to Figure 7.3 and applying the kinematic relation (3.32), the kinematic acceleration vector expressed in the inertial coordinate system is

$$\begin{bmatrix} A_X \\ A_Y \\ A_Z \end{bmatrix} = \ddot{\boldsymbol{L}}_I + R^T(\varphi, \vartheta, \psi)\,[[\omega]][[\omega]]\boldsymbol{H}_b +$$

$$+ R^T(\varphi, \vartheta, \psi)\,[[\dot{\omega}]]\boldsymbol{H}_b + \tag{7.6}$$

$$+ 2R^T(\varphi, \vartheta, \psi)\,[[\omega]]\dot{\boldsymbol{H}}_b + R^T(\varphi, \vartheta, \psi)\,\ddot{\boldsymbol{H}}_b$$

where $\boldsymbol{L}$ is the vector between the origins of the inertial and intermediate coordinate frames and $\boldsymbol{H}$ is the vector between the origins of the intermediate and body reference frames. The subscripts I and b underline that the vectors are expressed in the inertial and body reference frame respectively. The expression of the accelerations in Equation (7.6) represents the formulation of the signals that is implemented in the Bikesim simulator.

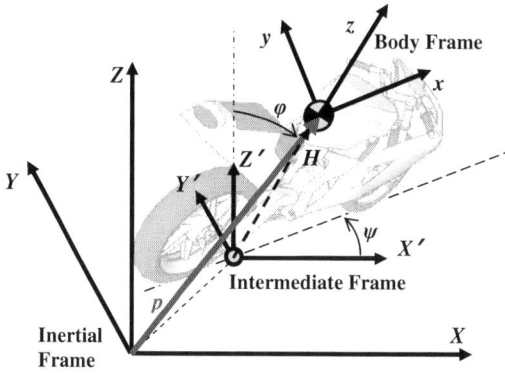

Figure 7.3: Definition of the vector that describe the position the center of mass of a motorcycle.

Even if it is the expression of the kinematic accelerations without any approximation, Equation (7.6) is not particularly useful for designing estimation algorithm, thus, it can been simplified. The first considered hypothesis is that the instantaneous center of rotation of the pitch dynamic is the center of mass of the motorcycle. Indeed, the vector p in Figure 7.3 (position vector of the COG of the vehicle in the inertial reference frame), can be expressed in the inertial reference frame as

$$p = \begin{bmatrix} L_X + Hs_\varphi s_\psi \\ L_Y - Hs_\varphi c_\psi \\ L_Z + Hc_\varphi \end{bmatrix} \tag{7.7}$$

where H is the module of the vector $\boldsymbol{H}$ (height of the center of mass), L_X is the projection of the vector $\boldsymbol{L}$ on the X axis, L_Y is the projection of the vector $\boldsymbol{L}$ on the Y axis and L_Z is the projection of the vector $\boldsymbol{L}$ on the Z axis.

Differentiating Equation (7.7), the following expression is yield

$$\dot{p} = \begin{bmatrix} \dot{L}_X + \dot{H}s_\varphi s_\psi + H\dot{\varphi}c_\varphi s_\psi + H\dot{\psi}s_\varphi c_\psi \\ \dot{L}_Y - \dot{H}s_\varphi c_\psi - H\dot{\varphi}c_\varphi c_\psi + H\dot{\psi}s_\varphi s_\psi \\ \dot{L}_Z + \dot{H}c_\varphi - H\dot{\varphi}s_\varphi \end{bmatrix} \tag{7.8}$$

where $\dot{p}$ is the velocity vector of the point COG of the vehicle in the inertial coordinate system.

The accelerations of the center of mass of the motorcycle expressed in the inertial frame are obtained as the second derivative of p [79] as

$$\begin{bmatrix} A_X \\ A_Y \\ A_Z \end{bmatrix} = \begin{bmatrix} \ddot{L}_X + s_\varphi s_\psi \ddot{H} + 2c_\varphi s_\psi \dot{H}\dot{\varphi} + 2s_\varphi c_\psi \dot{H}\dot{\psi} + \\ \ddot{L}_Y - s_\varphi s_\psi \ddot{H} - 2c_\varphi c_\psi \dot{H}\dot{\varphi} + 2s_\varphi s_\psi \dot{H}\dot{\psi} + \\ \ddot{L}_Z + c_\varphi \ddot{H} - 2s_\varphi \dot{H}\dot{\varphi} + \\[4pt] +c_\varphi s_\psi H\ddot{\varphi} - s_\varphi s_\psi H\dot{\varphi}^2 + 2c_\varphi c_\psi H\dot{\varphi}\dot{\psi} + s_\varphi c_\psi H\ddot{\psi} - s_\varphi s_\psi H\dot{\psi}^2 \\ -c_\varphi c_\psi H\ddot{\varphi} + s_\varphi c_\psi H\dot{\varphi}^2 + 2c_\varphi s_\psi H\dot{\varphi}\dot{\psi} + s_\varphi s_\psi H\ddot{\psi} + s_\varphi c_\psi H\dot{\psi}^2 \\ -c_\varphi H\ddot{\varphi} + s_\varphi H\dot{\varphi}^2 \end{bmatrix} \tag{7.9}$$

Even if an important simplification has been posed, Equation (7.9) is not easy to interpret and it is unnecessarily complicated.

A simplified formulation of the kinematic accelerations can be deduced considering that the mounting point of the IMU is the CIR (Centre of Instantaneous Rotation) of the roll and pitch dynamic; consequently, the principal terms that affect the kinematic accelerations in a two-wheeled vehicle are the longitudinal acceleration ($\dot{V}_x$), the centrifugal acceleration ($\dot{\psi}N_x$), the lateral acceleration ($\dot{V}_y$) and the vertical acceleration ($\dot{V}_z$). All of these contributions can be expressed in a reference frame that is rotated by an angle ψ around the absolute Z axis with respect to the inertial (or intermediate) frame; consequently, Equation (7.10) is yield.

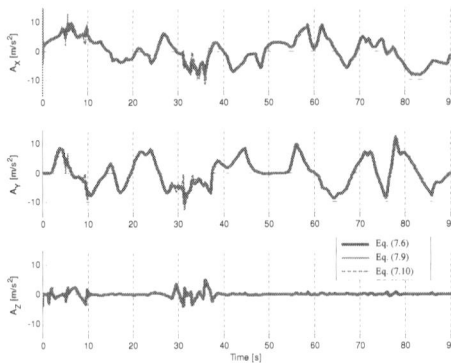

Figure 7.4: Comparison of the kinematic acceleration defined by Equation (7.6) (blue line), Equation (7.9) (green line) and Equation (7.10) (dashed red line).

$$
\begin{bmatrix} A_x \\ A_y \\ A_z \end{bmatrix} = R_Z^T(\psi) \begin{bmatrix} \dot{V}_x \\ \dot{\psi} V_x + \dot{V}_y \\ \dot{V}_z \end{bmatrix}
\tag{7.10}
$$

The models (7.9) and (7.10) are compared to formulation of the kinematic accelerations in Equation (7.6) using MSC BikeSim® simulator: the results are shown in Figure 7.4. The accelerations model (7.10) catches the principal terms that define the kinematic acceleration vector of a two-wheeled vehicle.

To take into account the contribution of the gravitational acceleration, it has to be considered that the acceleration measurement provided by an accelerometer is based on the measurement of a force [84]. Thus, an accelerometer sense the force that prevents it from falling toward the center of the Earth and it means that the device measures a positive gravitational acceleration. Consequently, the following model is yield:

$$
\begin{bmatrix} a_x \\ a_y \\ a_z \end{bmatrix} = R_{ZXY}(\varphi, \vartheta, \psi) \begin{bmatrix} A_X \\ A_Y \\ A_Z + g \end{bmatrix}
\tag{7.11}
$$

where $[0 \; 0 \; g]^T$ is the gravitational acceleration expressed in the inertial reference frame.

If model (7.10) is considered in Equation (7.11), the accelerations measured on the COG of the vehicle in a noiseless condition are

$$
\begin{bmatrix} a_x \\ a_y \\ a_z \end{bmatrix} = \begin{bmatrix} -c_\varphi s_\vartheta(\dot{V}_z + g) + c_\vartheta \dot{V}_x + s_\varphi s_\vartheta(\dot{\psi} V_x + \dot{V}_y) \\ s_\varphi(\dot{V}_z + g) + c_\varphi(\dot{\psi} V_x + \dot{V}_y) \\ c_\varphi c_\vartheta(\dot{V}_z + g) + s_\vartheta \dot{V}_x - s_\varphi c_\vartheta(\dot{\psi} V_x + \dot{V}_y) \end{bmatrix}.
\tag{7.12}
$$

In Figure 7.5, the accelerations model (7.12) is compared with the signals obtained by the virtual sensors of the simulator and it is noticeable that it provides a nice description of the noiseless acquired signals.

The same model (7.3)-(7.4) that has been adopted to describe the noise of a gyroscope can be adopted to characterize the noise of an accelerometer [85]:

$$
\tilde{a}(t) = a(t) + \Delta_a(t) + \eta_a(t)
$$
$$
\dot{\Delta}_a(t) = \eta_{\Delta_a}(t)
$$

$$
\tag{7.13}
$$

$$
E[\eta_a(t)\eta_a^T(\tau)] = \sigma_a^2 \delta(t - \tau)
$$
$$
E[\eta_{\Delta_a}(t)\eta_{\Delta_a}^T(\tau)] = \sigma_{\Delta_a}^2 \delta(t - \tau)
$$

where $\tilde{a}(t)$ is the continuous-time measured angular rate, $\Delta_a(t)$ is the model of the offset of the accelerometer and $\eta_a(t)$ and $\eta_{\Delta_a}(t)$ are independent zero mean Gaussian white noise process with variance σ_a^2 and $\sigma_{\Delta_a}^2$ respectively.

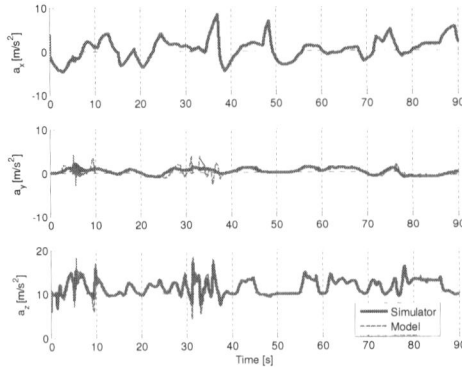

Figure 7.5: Validation of the model in Equation (7.12) of the measured accelerations.

Some final observations on the models of the acquired inertial sensors can be done:

- *Pitch and roll estimation*: the inertial measurements are dependent on roll and pitch, then, using these information an estimation of the roll angle compensating the effect of the pitch dynamic would be feasible;

- *Slope and bank*: the main drawback of the information provided by the inertial sensors is constituted by the presence of banks. On one hand, with the measured angular rates and accelerations the pitch Euler angle could be estimated: as stated by Equation (**4.1**) this quantity is also informative of the road slope α, therefore if the pitch angle is compensated for estimating the roll angle also the effect of a road slope will be cancelled out. On the contrary, estimating the roll Euler angle, the contribution of the bank of the road β and of the lean angle φ_R cannot be separated, this results in an additive error equal to the bank between the lean angle estimated via inertial sensors and road lean angle;

- *Offset compensation*: to employ the signals provided by the MEMS inertial sensors is necessary to consider a procedure to remove the bias of the sensors.

7.2 Design of an inclinometer

In the previous Section, the models of the inertial measurements in the body coordinate system have been developed. In a motorcycle application it cannot be guarantee that the mounting attitude of the IMU is aligned with the body frame. This problem can be overcome adopting the signals acquired by the accelerometers to estimate the mounting attitude of the inertial platform.

It's easy to note that the angular rates are dynamical measures (they may assume a non-null value only if the speed of the vehicle is greater than zero), while the accelerations are composed by dynamical components (longitudinal and centrifugal acceleration) and a static component (gravitational acceleration). The most important consequence is that the angular rates cannot be used to estimate the static roll angle, while it is well known that the acceleration's measurements are widely adopted to realize electronic inclinometer that measure the orientation of the gravitational acceleration [80], [86], [11], [87].

In this Section an optimal method to compute the static attitude of the inertial unit is described. Referring to Equations (7.12) and (7.13), the signals measured by the accelerometers in a static (null speed of the vehicle) and noiseless condition are described by

$$
\begin{bmatrix} a_x \\ a_y \\ a_z \end{bmatrix} = \begin{bmatrix} -c_\varphi s_\vartheta g \\ s_\varphi g \\ c_\varphi c_\vartheta g \end{bmatrix},
\tag{7.14}
$$

thus

$$
a_x^2 + a_y^2 + a_z^2 = g^2.
\tag{7.15}
$$

Firstly, it has to be noticed that the expression in Equation (7.14) is not a function of the yaw angle ψ, consequently:

- with the available signals just the roll and pitch angle can be recovered
- the static yaw angle ψ_o has to known a priori (to compute also the mounting yaw angle a fourth accelerometer is necessary).

To compute the static orientation of the inertial platform, the function cost J_g defined in Equation (7.16) is considered: the optimum values of the roll and pitch angle are such that J_g is minimized.

$$
J_g(\varphi, \vartheta) = \left(\tilde{a}_x + c_\varphi c_\vartheta g \right)^2 + \left(\tilde{a}_y - s_\varphi g \right)^2 + \left(\tilde{a}_z - c_\varphi c_\vartheta g \right)^2
$$

$$
\varphi_o, \vartheta_o : \arg\min_{\varphi, \vartheta} \left[J_g(\varphi, \vartheta) \right]
\tag{7.16}
$$

The solutions of the problem posed in Equation (7.16) are such that:

$$
\begin{cases} \dfrac{\partial J(\varphi, \vartheta)}{\partial \varphi} = 0 \\ \dfrac{\partial J(\varphi, \vartheta)}{\partial \vartheta} = 0 \end{cases}
\tag{7.17}
$$

thus

$$\begin{cases} -s_\varphi s_\vartheta \tilde{a}_x - c_\varphi \tilde{a}_y + s_\varphi c_\vartheta \tilde{a}_z = 0 \\ c_\varphi c_\vartheta \tilde{a}_x + c_\varphi s_\vartheta \tilde{a}_z = 0 \end{cases} \tag{7.18}$$

and

$$\vartheta_o = \arctan\left(\frac{-\tilde{a}_x}{\tilde{a}_z}\right)$$

$$\varphi_0 = \arctan\left(\frac{\tilde{a}_y}{c_{\vartheta_o}\tilde{a}_z - s_{\vartheta_o}\tilde{a}_x}\right) \tag{7.19}$$

If the inertial signals have a static orientation represented by the angles $(\varphi_o, \vartheta_o, \psi_o)$, the measurements in the body reference frame are computed as

$$\begin{bmatrix} \tilde{a}_x \\ \tilde{a}_y \\ \tilde{a}_z \end{bmatrix} = R_{ZXY}^T(\varphi_o, \vartheta_o, \psi_o) \begin{bmatrix} \tilde{a}_{x,o} \\ \tilde{a}_{y,o} \\ \tilde{a}_{z,o} \end{bmatrix}$$

$$\begin{bmatrix} \tilde{\omega}_x \\ \tilde{\omega}_y \\ \tilde{\omega}_z \end{bmatrix} = R_{ZXY}^T(\varphi_o, \vartheta_o, \psi_o) \begin{bmatrix} \tilde{\omega}_{x,o} \\ \tilde{\omega}_{y,o} \\ \tilde{\omega}_{z,o} \end{bmatrix}, \tag{7.20}$$

where $(\tilde{a}_{x,o}, \tilde{a}_{y,o}, \tilde{a}_{z,o})$ and $(\tilde{\omega}_{x,o}, \tilde{\omega}_{y,o}, \tilde{\omega}_{z,o})$ are the inertial measurements in the reference frame rotated of $(\varphi_o, \vartheta_o, \psi_o)$ with respect to the body coordinate system.

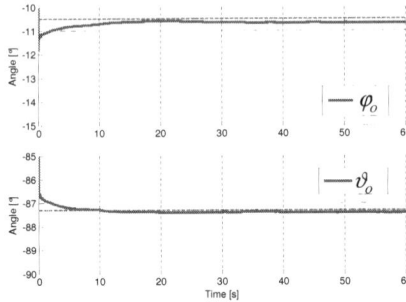

Figure 7.6: Comparison between the reference mounting attitude of the IMU (dashed red line) and the output of the estimation algorithm (blue line).

Equation (7.19) has to be applied considering just the mean of the interesting signals and that the lean angle of the motorcycle is null. In Figure 7.6 the estimation result for a test in Imola is depicted (the dashed red line represents the mounting attitude geometrically measured knowing the motorcycle characteristics). The data have been acquired positioning the motorbike on front and rear stands for one minute.

The signals aligned in the body coordinate system are adopted to estimate the offset of the inertial signals. Consequently, it can be noticed that the estimated offset are the offset of the aligned signals and not the offset of the acquired inertial measurements.

The described flow of the data is necessary mainly because the unbiasing procedure also compensate the effect of the errors that affect the estimation of the angles φ_o, ϑ_o.

7.3 Estimation of the offset of the inertial sensors

In this Section the estimation of the offset of the inertial sensors is taken into account. In the attitude estimation context, this problem is commonly addressed by correcting the inertial measurement errors with other measurement systems such as GPS, gravity sensors, external speed sensor, barometric sensor and magnetic compass [13], [88], [89], [90]. The corrections provided by the fusion of different sources of data are typically based on estimation via Kalman Filter (see Section 3.2).

In the experimental set-up described in Section 4.3.2, signals for the correction of the measurement errors are not available, moreover the ECU mounted in a modern motorbike have a limited computational capacity, then the implementation of an high order Kalman Filter for the offset estimation is not feasible in a real time application. Consequently in this subsection a buffer-based algorithm for offset estimation is presented. Nevertheless, in Section 9.1.3.1 it will be shown how to estimate the measurement errors in Kalman Filter framework using just inertial signals.

It has to be mentioned that the signals that are adopted in this application are temperature compensated on-line using data provided by static characterization of the inertial unit. The calibration of the unit in terms of gain and nominal bias has been performed with well known procedures such as [91], [42], [92] and [93].

The drift-over-time has to be estimated. In this Section, the problem of offset estimation is approached with a grey box methods that is based both on the models described in the previous Section and on the analysis of the condition in which the vehicle is operating.

The problem of offset estimation of the inertial signals acquired on a motorbike is further complicated by the fact that:

- Due to the correlation between the roll and yaw dynamics (Equation (7.2)), the signal ω_y does not oscillate around the null value (Figure 7.7a), in fact for right turn $\varphi > 0$ and $\dot{\psi} < 0$, then $s_\varphi \dot{\psi} < 0$, and for left turn $\varphi < 0$ and $\dot{\psi} > 0$, then $s_\varphi \dot{\psi} < 0$;

- Due to the relation between the roll dynamic, the centrifugal acceleration and the gravitational acceleration (Equation (7.12)) a_z do not oscillate around the null value (Figure 7.7b), in fact $a_z \geq g$ both for right and left turn.

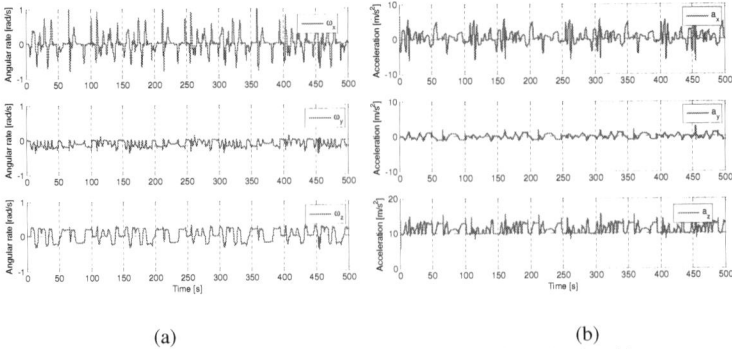

(a)   (b)

Figure 7.7: Plot of angular rates (a) and accelerations (b) for a simulated long run.

It is necessary to define:

- which are the conditions in which the mean of the inertial signals is null
- auxiliary signals such that their mean is null.

Taking into account the condition in which the acquired signal should be null, it is trivial to note that:

- the angular rates are null if the attitude of the vehicle in a steady state condition (Equation (7.2))
- if the vehicle has null roll angle and null pitch angle then $a_x \cong \dot{V}_x$, $a_y \cong \dot{V}_y$ and $a_z \cong g + \dot{V}_z$ (Equation (7.12)), and considering that for a motorcycle with null attitude the quantities $\dot{V}_y$ and $\dot{V}_z$ can be neglected [94], $a_x \cong \dot{V}_x$, $a_y \cong 0$ and $a_z \cong g$.

As a consequence of the previous considerations, the following auxiliary variables can be defined:

$$\gamma_x = a_x - \dot{V}_x$$
$$\gamma_z = a_z - g$$

(7.21)

It can be concluded that:

1. the offset of the gyroscopes signals can be studied when the speed of the vehicle is null
2. the offset of the accelerometers can be analyzed when the signals provided by the gyroscopes are equal to zero; in particular the offset of a_x is equal to the offset of γ_x and the offset of a_z is equal to the offset of γ_z.

The important aspect of the previous points is that they are independent in the sense that the offset of the gyroscopes and accelerometers can be studied in two uncorrelated conditions. The drawback is that the condition that defines when the offset of the accelerometers can be estimated depends on the estimation output of angular rate offsets observer.

It is well known [84], [95] that the offsets introduced in Equations (7.3) and (7.13) are slowly varying quantities, thus, before an offset observer is applied, the interesting signal are filtered at a frequency of 0.1 Hz and down-sampled at 1 Hz. The signals so obtained are used on a estimation algorithm based on two circular buffers of different length ΔT_1 and ΔT_2 with $\Delta T_2 > \Delta T_1$ as depicted in Figure 7.8, in which:

- C_1 *(test)*: it is verified if the acquired signal is in a condition in which the mean should be null;
- C_2 *(test)*: it is verified if the buffer of length ΔT_1 is full;
- C_3 *(test)*: it is verified if the buffer of length ΔT_2 is full;
- *Offset update*: the offset is updated computing the mean of the values stored in the buffer of length ΔT_2.

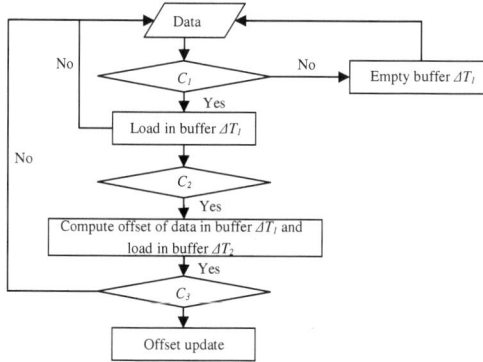

Figure 7.8: State flow of the algorithm with two circular buffers for the estimation of the offset of the inertial signals.

To compute the offset of the data stored in the buffer of length ΔT_1, the sample mean can be used.

In Figure 7.9a and Figure 7.9b it is depicted the offset estimation of the acceleration signals and angular rates signals respectively. The considered condition is a not moving motorcycle. In the example $\Delta T_1 = 10$ s and $\Delta T_2 = 60$ s are considered and it shown that the double buffer algorithms correctly estimate the offset of the inertial signals.

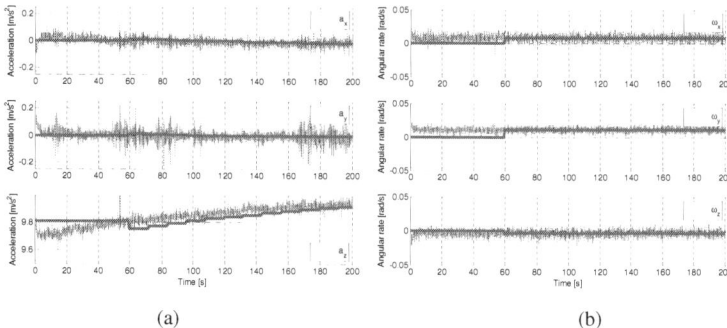

(a)   (b)

Figure 7.9: Example of offset estimation (red line) while the motorcycle is not moving. (a) Accelerations offset estimation. (b) Angular rates offset estimation.

7.4 Concluding remarks

In this Chapter it has been described how the problem of a attitude estimation has been approached and some preliminaries for the solution of the problem have been presented.

The models of the inertial signals have been discussed and in particular for the acquired accelerations three models have been compared.

The last two Sections have been devoted to the description of the algorithms to align the IMU to the body coordinate system and to compensate the offset of the acquired signals.

In what follows, it will be considered that the input signals of the estimation algorithm are aligned to the body frame and depurated of the offset.

Chapter 8
Motorcycle attitude estimation with inertial sensors: frequency split

This Chapter is devoted to the presentation of an innovative solution to address the problem of lean angle estimation of a motorcycle. The proposed algorithms are based on the frequency separation principle that is a frequency method to realize the data fusion of different sensors.

The principle is quite simple. In many applications, such the one considered herein, two kinds of measurements are available: one set of measurements is characterized by an high frequency noise and a second set of measurements that is affected by a low frequency noise. The first set of signals can be used to obtain information about the low frequency component of the quantity that would like to be estimated, while the high frequency component is recovered by the second set of signals.

This principle will be used to estimate the lean angle of the motorbike: from the roll gyroscope the high frequency component of the lean angle is recovered, while the low frequency component is estimated using the low frequency component of the acquired inertial signals. The estimation of the low frequency component of the lean angle is the main topic of the chapter.

The Chapter is organized as follows. Considering that the inputs of the attitude estimation algorithm are the inertial signals aligned in the body frame and unbiased (Figure 7.1), in Section 8.1 the frequency separation principle will be introduced and it will be shown that just the high frequency component of the lean angle can be estimated by the integration of the roll gyroscope. In Section 8.2, the low frequency estimation problem is analyzed in a Neural Network framework [21]. Finally, in Section 8.3, the low frequency estimation problem is addressed with a model base approach [19]. It will be shown that, even if this method has good estimation accuracy, it is based on inertial measurements that describe the attitude of the motorcycle in terms of Euler angles instead of road attitude angles, thus, additive errors due to banked roads cannot be compensated; moreover, the performance strictly depends on the pitch dynamic of the motorcycle.

8.1 The frequency separation principle

At a first glance, the problem of estimating the roll angle of a motorbike may appear trivial, using a standard inclinometer (Chapter 7). However, since the stability of a motorcycle is based on the equilibrium between gravitational and centrifugal forces [3], it is easy to understand that a standard body-fixed inclinometer is useless.

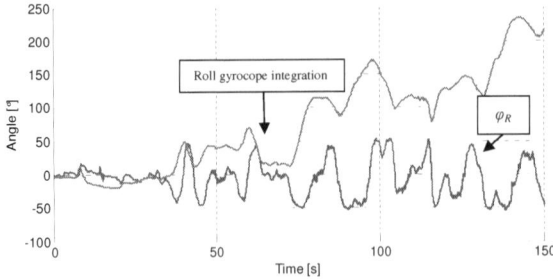

Figure 8.1: Real roll angle (φ_R, blue line) and roll angle estimated with a simple integration of the roll angular rate (green line).

A more sound way of measuring the roll angle is the numerical integration of the rotational speed along the roll axis, measured by a gyroscope. However, numerical integration is highly sensitive to DC measurement errors and it is well known that the noise of the inertial measurement produce an accumulative error, which induce a drift in the integrated signal (see Figure 8.1, where this simple approach is applied on a set of real measurements). This method can be used only if the integration procedure is used over short time windows, and a robust re-initialization procedure is available. In a motorcycle none of these two conditions hold, and this simple and naïve approach cannot be consistently used.

The estimation approach proposed herein has the general architecture shown in Figure 8.2 [19]. As mentioned above, the main idea is that of splitting the input measured inertial signals and speed into High-Frequency (subscript *HF*) and Low-Frequency (subscript *LF*) components. Such signal components are processed independently after having been split by the Frequency Separation block and then - at each sampling instant - the LF estimate $\hat{\varphi}_{LF}$ and the HF estimate $\hat{\varphi}_{HF}$ are added to build the final estimate $\hat{\varphi}$ of the roll angle.

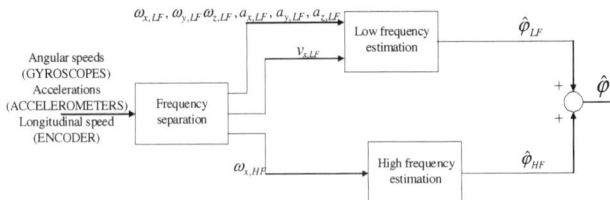

Figure 8.2: High-level architectural view of the proposed estimation algorithm.

The internal structure of the Frequency Separation block is illustrated in Figure 8.3. It is constituted by a standard first-order High Pass (HP) filter with a cutoff frequency f_{sp} and a summing junction which splits the signal into the LF and HF components. The parameter f_{sp} is the only tuning knob.

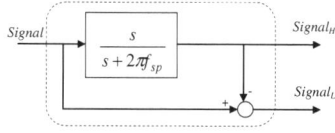

Figure 8.3: Internal structure of the Frequency Separator block.

Notice that the Low-Frequency estimation block receives as inputs the LF components of the measured body-fixed inertial signals, and the longitudinal speed v; instead, the High-Frequency estimation block only uses the HF component of the body-fixed roll rate ω_x.

Considering the expression of $\tilde{\omega}_x$ in Equation (7.3), the integration of the acquired signal yields:

$$\int \tilde{\omega}_x dt = \int c_{\vartheta} \dot{\varphi} dt - \int s_{\vartheta} c_{\varphi} \dot{\psi} dt + \int (\Delta_\omega + \eta_\omega) \tag{8.1}$$

and applying the rule of integration by parts

$$\int \tilde{\omega}_x dt = c_{\vartheta} \varphi + s_{\vartheta} \int \left(\varphi \dot{\vartheta} - c_{\varphi} \dot{\psi} \right) dt + \int (\Delta_\omega + \eta_\omega) dt =$$
$$= c_{\vartheta} \varphi + I_{\dot{\vartheta},\dot{\psi}} + I_{\Delta,\eta}$$
$$I_{\dot{\vartheta},\dot{\psi}} = s_{\vartheta} \int \left(\varphi \dot{\vartheta} - c_{\varphi} \dot{\psi} \right) dt \tag{8.2}$$
$$I_{\Delta,\eta} - \int (\Delta_\omega \mid \eta_\omega) dt$$

In Equation (8.2) the terms of error between φ and the integral of $\tilde{\omega}_x$ are highlighted:

- *Pitch angle*: the pitch dynamic of the vehicle introduce a multiplicative error $c_{\vartheta}\varphi$ and a null mean additive error $I_{\dot{\vartheta},\dot{\psi}}$;

- *Noise*: the integration of the bias and noise causes a cumulative error $I_{\Delta,\eta}$; notice that this term of error is not null even if the bias of the signal is compensated (the integrator of a white noise signal is not zero).

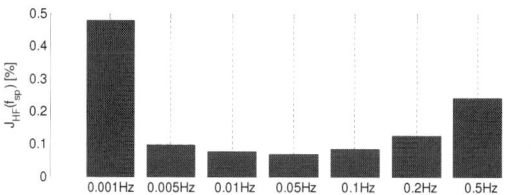

Figure 8.4: Analysis of the cost function J_{HF}.

The parameter f_{sp} of the frequency separation filter can be tuned to minimize the cost function

$$J_{HF}(f_{sp}) = \frac{mse(\varphi_{HF,f_{sp}} - \hat{\varphi}_{HF,f_{sp}})}{mse(\varphi_{HF,f_{sp}})} \tag{8.3}$$

where $\varphi_{HF,f_{sp}}$ is the HF component of the roll angle of the vehicle, $\hat{\varphi}_{HF,f_{sp}}$ is the high frequency component of the estimated attitude angle and J_{HF} is the Error-Signal-Ratio (ESR).

The cost function J_{HF} has been evaluated considering a long run in Misano and Imola circuits. In Figure 8.4 the mean performance $(J_{HF,Misano} + J_{HF,Imola})/2$ is depicted.

The sensitivity of J_{HF} to f_{sp} is not high, in particular if $f_{sp} \in [0.005,0.2]Hz$ the performance index is around 10%. In Figure 8.5 the results of a long run experiment in the Imola circuit are displayed considering $f_{sp} = 0.2\,Hz$. The estimated HF roll angle is compared with the HF true angle (measured with the optical sensors). Apparently, the high frequency estimation is very accurate and the drift problem is completely removed.

Using the algorithm in Figure 8.2, the measurement noise is depurated of the LF component, and the drift problem is removed. On the other hand, this algorithm also removes the LF component of the roll rate $\tilde{\omega}_x$: the estimated roll angle $\hat{\varphi}_{HF}$ hence carries no information on the LF component of the roll angle, which must be reconstructed following a different path.

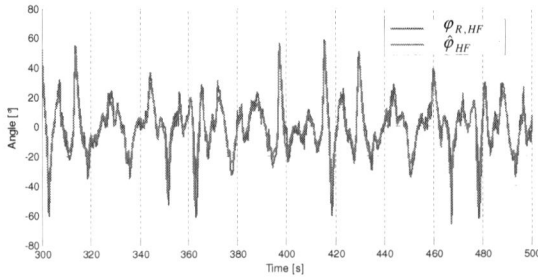

Figure 8.5: HF-components: true roll angle (blue line) and estimated roll angle (green line).

In the following the estimation of the LF component of the lean angle is carried out: first of all the performance of the performances of the algorithm in Figure 8.2 are studied for different sensors configurations in a Neural Network framework; secondly algorithms to estimate the low frequency component of the roll angle are presented in Section 8.3.

8.2 Performance analysis of frequency separation based algorithm

The aim of this Section is not to define the absolute best performance of the estimation of the lean angle with the frequency separation principle. The estimation performance strictly depends on the condition in which the inertial signals are acquired, thus to define the best performance a huge amount of data should be needed. In particular, the target of this Section is to:

- Analyze the principal measurements to consider for performing the estimation of the lean angle;
- Study the conditions that mainly afflict the estimation performance;
- Obtain indication on which are the principal information that have to be derived from the inertial signals to perform the roll angle estimation.

In a simulation environment some different conditions can easily be analyzed trying to generate data that are representative of a big spread of operating conditions. As a consequence the results of this Section will be based in particular on simulated data. Experimental data will be also analyzed, but being collected in a non-controllable environment, they will just used to validate the simulation results.

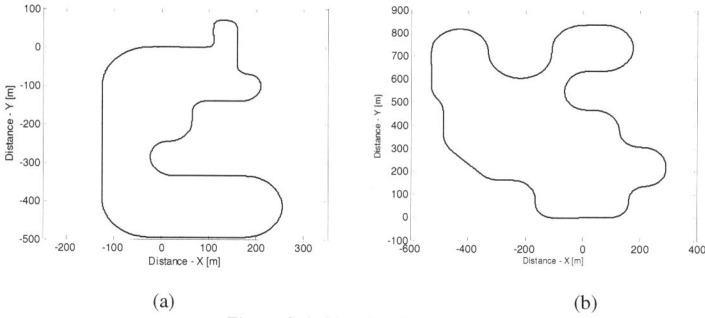

(a)   (b)

Figure 8.6: Simulated test tracks.

In the simulation environment the tracks in Figure 8.6 are build through segments definition. The tracks are composed by a sequence of turn with increasing radius from 20 m to 60 m (Figure 8.6a) and from 70 m to 150 m (Figure 8.6b), moreover the circuits are build with right and left curves of equal radius.ù

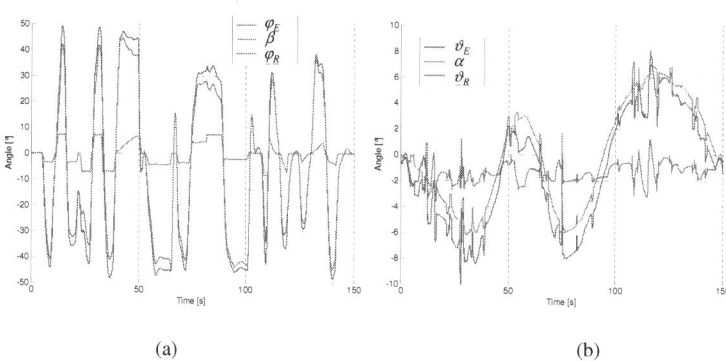

(a)   (b)

Figure 8.7: (a) Comparison between φ_E and φ_R with a banked road; (b) Comparison between ϑ_E and ϑ_R with a sloped road.

The attitude angles provided by the simulator are the inertial ones, then, if banks and slopes are simulated, the differences from the road attitude angles have to be computed as in Equation (4.1):

the reference tilt angle used in the following analysis is defined with respect to the road plane. In Figure 8.7 the comparison between the inertial and road attitude angles is depicted (track in Figure 8.6b simulated with slopes, banks and covered with variable speed).

8.2.1 Description of the adopted approach

First of all the information provided by the measurement axes on the roll angle estimation will be studied so that it will be underlined which are the fundamentals. Starting from this results, some particular configurations are compared in different condition using both the data provided by the simulator and the experimental test. All the results will be supported by the measurements models presented in Section 7.1.

The analysis is based on the frequency separation principle described in the previous Section. In Figure 8.8 the frequency separation scheme is recalled to highlight the static non-linear estimation block. On one hand, the estimation of the high frequency component of the roll angle is performed through the integration of the high frequency component of the roll gyroscope; on the other hand the low frequency component of the lean angle is estimated using a non linear statistical data modeling tool whose inputs are the low frequency components of different sub-sets of the inertial signals and the low frequency component of the speed signal. In particular, the estimation of the low frequency component of the lean angle can be performed with a feed-forward Neural Network [37], [38] for which different toolboxes are available.

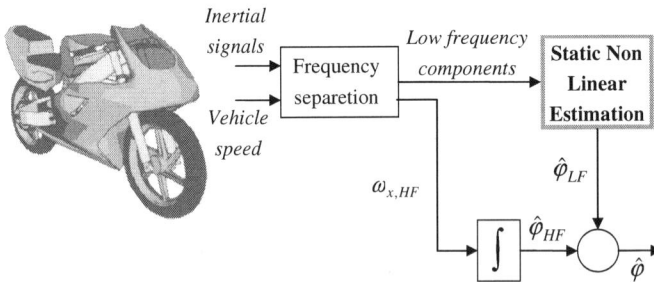

Figure 8.8: Adopted structure for the analysis of inertial signals contribution on the roll angle estimation.

In this application, the neural network is structured with only one hidden layer with activation functions based on symmetrical sigmoid and linear output layer; the number of unit of the hidden layer is chosen according to [49].

The data provided by the simulator and by the experimental test are adopted to conduct the analysis in two different conditions:

- *Simplified condition (on plane track covered at constant speed)*: in this framework the most important dynamic that affects the inertial signal is the roll angle while the pitch dynamic is almost constant;
- *General condition (track with slopes, banks and covered with variable speed)*: some dynamics other than the roll one are introduced; the purpose is to separately analyze some conditions that increase the error of the roll angle estimation.

For each condition the analysis is organized as follow:

- An optimization procedure is applied to the parameter of the filter (f_{sp}) that realize the frequency separation of the Neural Network input signals;
- Different sets of signals are compared;
- Conclusions are discussed pointing out the information that can be extracted from the considered set of signals and the principal estimation errors causes.

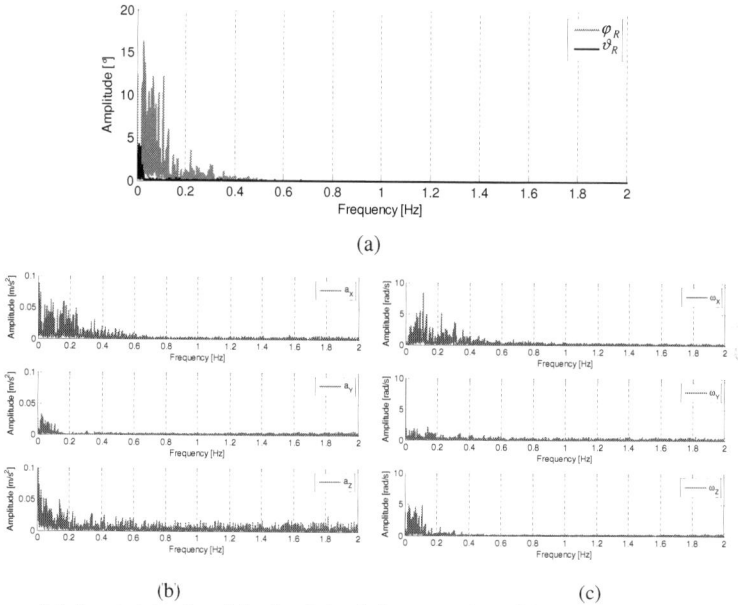

(a)

(b)   (c)

Figure 8.9: Spectral density of simulated signals in a general condition. (a) Attitude angles. (b) Accelerations. (c) Angular rates.

The value of f_{sp} has a remarkable role in the performance analysis. In Figure 8.9 the spectral densities of the attitude and inertial signals are depicted (signals generated in a simulated environment covering track in Figure 8.6b in a general condition). By comparing the picture depicted, it can be noticed that on one side there is a strong correlation between the spectrum of a_y and ω_z and the spectrum of the roll dynamic, on the other side the other signals have a larger bandwidth than the roll angle and the contribution of the roll dynamic is not plain to understand. The optimum value of f_{sp} is an indicator of the limit frequency at which the roll dynamic is the most important contribution on the measured inertial variables taken into account, or, in other words, which is the spectral content of the roll dynamic on the considered measurements.

To optimize the value of f_{sp}, the functional cost $J_r\left(\hat{\xi}, f_{sp}, S_I\right)$ defined in (8.4) is adopted (this is ESR of the estimation error), where $\hat{\xi}$ represents the vector of the parameters of the identified Neural Network and S_I is the set of signals used as input. J_r is the ratio between the Mean Square Error (MSE) of the identification error and the MSE of the reference signal. By this definition, the considered functional cost is normalized with respect to the amplitude of the reference variable.

$$J_r\left(\hat{\xi},f_{sp},S_I\right)=\frac{mse\left(\varphi_{R,LF,f_{sp}}-\hat{\varphi}_{LF,f_{sp}}\right)}{mse\left(\varphi_{R,LF,f_{sp}}\right)} \tag{8.4}$$

In Equation (8.4), $\varphi_{LF,f_{sp}}$ ($\hat{\varphi}_{LF,f_{sp}}$) is the low frequency component of the reference (estimated) lean angle applying a frequency separation filter with a cut-off frequency equal to f_{sp}. The tie between J_r and f_{sp} just depend on the contribution of the roll dynamic to the inertial measurement and do not depend on the amplitude of the roll angle within the bandwidth f_{sp}.

After the frequency analysis, the contribution of each measure to the lean angle estimation is studied. The available signals to identify the tilt angle are three accelerations, three angular rates and the vehicle speed. A signal representative of the speed of the vehicle is always available on a motorcycle, therefore, in the following the vehicle speed signal is always considered in the set of signals adopted for the estimate of the lean angle. The research will point out which of the inertial measurements are significant for the estimate of the lean angle of the motorcycle. Thank to the results of this analysis, some particularly important sets un signals are highlighted and tested in different conditions with the data provided by the simulator (flat track, banked road, sloped road and non-constant speed) and experimental test: this study is helpful to discover what information have to be recovered from the acquired inertial signals to optimize the roll angle estimation through frequency separation principle. The performance are compared recurring to the functional cost defined by

$$J_c\left(\hat{\xi},f_{sp},S_I\right)=\frac{J_r\left(\hat{\xi},f_{sp},S_I\right)}{J_r\left(\hat{\xi},f_{sp},S_{opt}\right)} \tag{8.5}$$

where the set of signals S_{opt} represents the one that gives the best performance among the compared sets of inputs to the Neural Network (the set S_{opt} is always the richest in terms of number of signals). Thanks to this definition the adopted functional cost is useful to describe the reduction of the performance with respect to the best one obtained with the set S_{opt} as input to the Neural Network. The performance indexes $J_r\left(\hat{\xi},f_{sp},S_I\right)$ and $J_c\left(\hat{\xi},f_{sp},S_I\right)$ are adopted because the principal aim is not to underline the absolute performance of the estimation of the roll angle via inertial sensors, but to highlight the decrease of the performance considering different environments and different configurations.

In the following Sections the described analysis is performed for the ideal and general condition.

8.2.2 Performance analysis in simplified condition with simulated data

In this Section the performance analysis is executed considering two circuit (Figure 8.6) covered at a constant speed and without considering the presence of banks and slopes so that the roll dynamic results the principal contribution that influence the measured inertial signal. The track in Figure 8.6a is covered with a speed from 20 km/h to 35 °km/h while the track in Figure 8.6b is covered with a speed from 40 km/h to 80°km/h. The minimum speed is defined by the constraints of the simulator (the multi-body model can be solved if the motorbike has a speed greater than 20 km/h), while the maximum is defined by the maximum speed at which the turn with the minimum radius curve can be approached. To the aim of guarantee an unbiased set of data for training the network (the means of the interesting signals have to be null), the tracks are covered both in clockwise sense and counterclockwise sense.

In Figure 8.10 the simulated attitude angles are depicted for the track in Figure 8.6 covered at 80 km/h. It can be notice that the pitch dynamic is almost null, but not exactly zero, this pitch dynamic has to be considered to correctly interpret the following performance analysis.

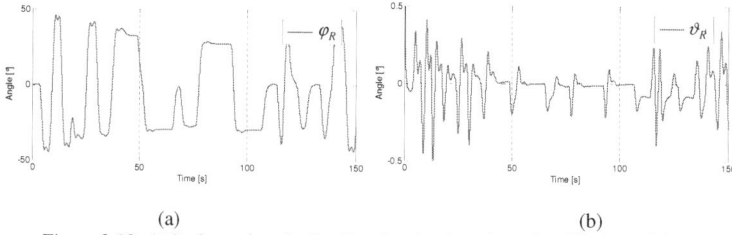

(a)   (b)

Figure 8.10: Attitude angles obtained by the simulator in a simplified condition.

Using the data set described above, the identification of the low frequency component of the lean angle is considered. First of all, considering the angular rates set (W) and the accelerations set (A) separately, the performance indexes $J_r\left(\hat{\xi}, f_{sp}, W\right)$ and $J_r\left(\hat{\xi}, f_{sp}, A\right)$ are compared for different values of the bandwidth of the frequency separation filter. Then, to go deep in the investigation, the contribution of each measurement axis is studied investigating the performance index $J_c\left(\hat{\xi}, f_{sp}, S_I\right)$ for different sets of input signals S_I. Finally, some mixed configurations are defined. These are particularly important both for designing innovative algorithms and for arranging the experimental set-up. In the case of mixed configurations of inputs set, the optimum value of f_{sp} is selected by the analysis of $J_r\left(\hat{\xi}, f_{sp}, F\right)$, where F represents the set of inputs composed by all the available inertial signals.

8.2.2.1 Analysis of the frequency contribution

It is well known that the roll dynamic has a bandwidth of about [0.5-1] Hz (Figure 8.9a). Due to the other dynamics of the motorcycle (pitch, longitudinal acceleration, vertical acceleration, etc.) and to the road conditions (slopes and banks), the roll dynamic has the most relevant contribution on the inertial signals just until a frequency smaller than the overall roll bandwidth: f_{sp} is an approximation of this limit frequency for the considered set of signals.

In Figure 8.11, the performance indexes $J_r\left(\hat{\xi}, f_{sp}, W\right)$ and $J_r\left(\hat{\xi}, f_{sp}, A\right)$ are compared for different values of f_{sp}. While the best performance of the accelerations set is reached at a frequency of 0.01 Hz, the best performance of the angular rates set is reached with a separation frequency of 0.2 Hz. Recalling Equations (7.2) and (7.12), the angular rate measures are informative just of the rotational motorcycle dynamic while the acceleration measures are affected both by translational and rotational contribution and by the constant gravitational acceleration. The main consequence is that the set of input W are more informative of the roll angle dynamic with respect to the input set A in which the roll dynamic is overlapped with secondary dynamics. The performance in the estimation of the low frequency component of the roll angle is better if the considered input signals are angular rates. This general result is due to presence of translational dynamics and secondary rotational dynamics that affect the measured accelerations.

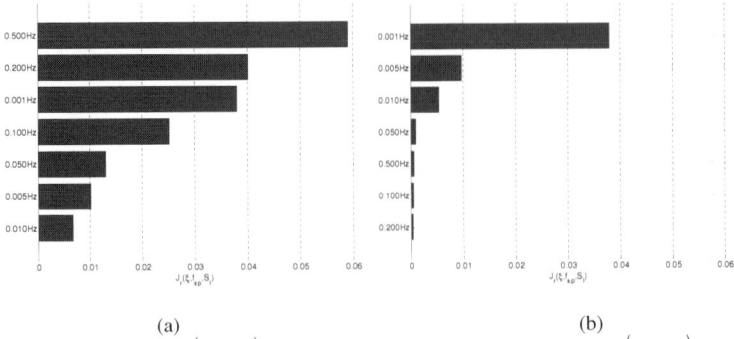

(a)   (b)

Figure 8.11: (a) Analysis of $J_r\!\left(\hat{\xi}, f_{sp}, A\right)$ for different values of f_{sp}. (b) Analysis of $J_r\!\left(\hat{\xi}, f_{sp}, W\right)$ for different values of f_{sp}.

8.2.2.2 Principal measurements axes

The contribution of each acquired signals is analyzed. It is well known that the position and orientation of a rigid body can be described by 6 Degrees of Freedom (DOF) that are measured by the adopted inertial sensors. The question that would be useful to be answered to is which is the fundamental subset of measurements to perform an estimation of the roll angle of a motorbike.

First of all the models of the acquired signals have to be studied. Consider the on plane condition with constant speed, consequently, the pitch dynamic is almost null ($\vartheta(\cdot) \cong 0$) and the lateral speed $\dot{V}_y$ can be neglected, then, Equations (7.2) and (7.12) can be simplified as

$$
\begin{aligned}
\omega_x &\cong \dot{\varphi} & a_x &\cong \dot{V}_x \\
\omega_y &\cong \sin(\varphi)\dot{\psi} & a_y &\cong -g\sin(\varphi) - \dot{\psi}V_x\cos(\varphi). \\
\omega_z &\cong \cos(\varphi)\dot{\psi} & a_z &\cong g\cos(\varphi) + \dot{\psi}V_x\sin(\varphi)
\end{aligned}
\tag{8.6}
$$

Observing Equation (8.6), some comments can be done:

- The angular rate ω_x does not depend on the roll angle φ.

- The angular rates ω_y and ω_z are a non linear function of the roll angle and, for typical lean angle range of $\pm 60°$, the principal contribution is measured on the yaw axis. This outcome can be explained noticing that if $\varphi < 45°$ then $\omega_y/\omega_z < 1$ and if $\varphi > 45°$ then $\omega_y/\omega_z > 1$; moreover the sign of ω_z is correlated with the sign of the lean angle, while the sign of the lateral angular rate measured in the body frame does not depend on the roll dynamic (the roll and yaw dynamic of the motorcycle are opposite in sign, then, the sign of the lateral angular rate is always negative and the sign of the vertical one is informative of the sign of the roll dynamic).

• The measured acceleration a_x does not depend on the lean angle and is strongly related to longitudinal acceleration of the vehicle.

• For the typical lean angle range a_y is characterized by higher dynamical contribution (centrifugal acceleration projected on the considered axis) and higher static sensitivity (sensitivity to the gravitational acceleration) with respect to a_z.

Thanks to the previous comments, it is expected that the two fundamental variables to perform the estimation of the roll angle are the angular rate measured around the yaw axis ω_z and the lateral acceleration a_y. This result is confirmed by the bar plot depicted in Figure 8.12. In this Figure the performance index $J_c\left(\hat{\xi}, f_{sp}, S_I\right)$ is analyzed to highlight the loss of information due to the removal of a signal from the set of inputs to the Neural Network.

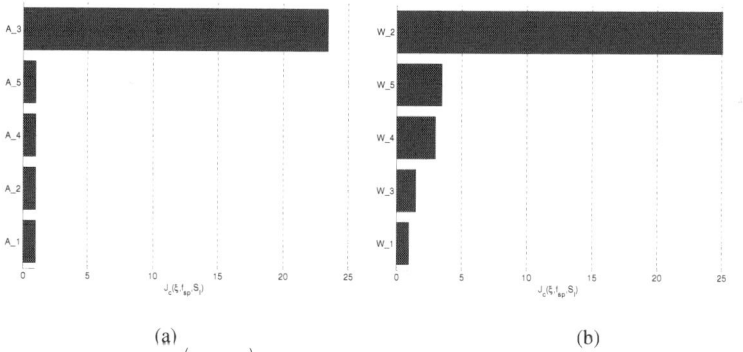

(a)   (b)

Figure 8.12: (a) Analysis of $J_c\left(\hat{\xi}, f_{sp}, S_I\right)$ for different sets of accelerations and $f_{sp}=0.01$ Hz. (b) Analysis of $J_c\left(\hat{\xi}, f_{sp}, S_I\right)$ for different sets of angular rates and $f_{sp}=0.2$ Hz.

In Figure 8.12a the performance index $J_c\left(\hat{\xi}, f_{sp}, S_I\right)$ is compared considering $f_{sp}=0.01$ Hz, set A_I as the best set S_{opt} ($S_{opt}=A_I$) and:

• Set A_1: $\{a_x, a_y, a_z, V_x\}$

• Set A_2: $\{a_x, a_y, V_x\}$

• Set A_3: $\{a_x, a_z, V_x\}$

• Set A_4: $\{a_y, a_z, V_x\}$

• Set A_5: $\{a_y, V_x\}$

In Figure 8.12b the comparison of the performance $J_c\left(\hat{\xi}, f_{sp}, S_I\right)$ is conducted considering $f_{sp}=0.2$ Hz, $S_{opt}=W_I$ and:

• Set W_1: $\{\omega_x, \omega_y, \omega_z, V_x\}$

• Set W_2: $\{\omega_x, \omega_y, V_x\}$

- Set W_3: $\{\omega_x, \omega_z, V_x\}$
- Set W_4: $\{\omega_y, \omega_z, V_x\}$
- Set W_5: $\{\omega_z, V_x\}$

The drop off of the performance that can be observed in $J_c(\hat{\xi}, f_{sp}, A_3)$ and $J_c(\hat{\xi}, f_{sp}, W_2)$ underlines the importance of the measurement of the lateral acceleration and relative yaw rate.

The bar plot depicted in Figure 8.12 also shows that neglecting the low frequency component of the angular rate ω_x, a minor drop off of the performance is introduced. The roll gyroscope bring information about the small pitch dynamic depicted in Figure 8.10b: this information is needed to correctly estimate the roll angle of the motorbike as described in Section 7.1.

In the following, the performance of mixed configurations is taken into account. The compared sets are defined starting from the previous results. The following inputs sets are then considered:

- Set F: $\{a_x, a_y, a_z, \omega_x, \omega_y, \omega_z, V_x\}$
- Set AW_1: $\{a_y, a_z, \omega_y, \omega_z, V_x\}$
- Set AW_2: $\{a_y, \omega_z, V_x\}$

and the performance index $J_c(\hat{\xi}, f_{sp}, S_I)$ is compared considering $S_{opt} = F$.

8.2.2.3 Performance of the lean angle estimate with data fusion

If the set of input signals is constituted by a accelerations and angular rates, a new analysis of the bandwidth of the frequency separation filter has to be conducted.

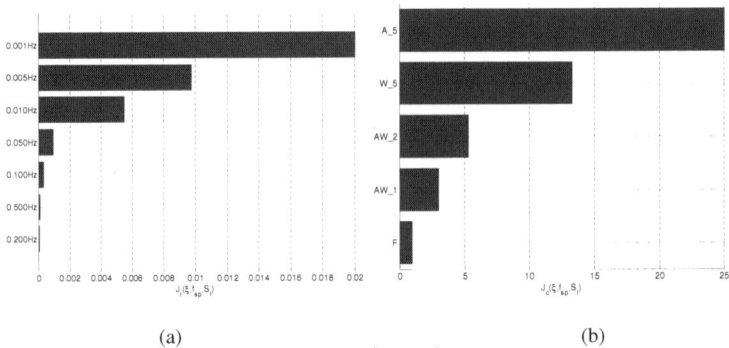

(a)   (b)

Figure 8.13: (a) Analysis of the performance index $J_r(\hat{\xi}, f_{sp}, S_I)$ for a full sensors configuration. (b) Analysis of $J_c(\hat{\xi}, f_{sp}, S_I)$ for mixed input configurations in simplified condition.

In Figure 8.13a the comparison of $J_r(\hat{\xi}, f_{sp}, S_I)$ is depicted for different values of f_{sp}. The optimum performance index is reached with $f_{sp} = 0.2\ Hz$. This value is due to the information provided by the angular rates that have a bandwidth of influence of the roll dynamic greater than the one that characterize the acceleration signals.

In Figure 8.13b the performance index $J_c\left(\hat{\xi}, f_{sp}, S_I\right)$ is compared considering $S_{opt}=F$ and the input sets F, AW_1, AW_2, W_5 and A_5. The principal result of this comparison is the significance of the information bring by the angular rate measured around the yaw axis and the lateral acceleration. It is also confirmed the contribution due to the signals measured along the longitudinal axis.

In the following Section a more general environment is taken into account. Adopting simulated data, it is considered the presence of slopes and banks on the road, moreover the tracks depicted in Figure 8.6 are covered at a variable speed.

8.2.3 Performance analysis in general condition with simulated data

From the analysis conducted in Section 8.2.2, the importance of the information bring by the angular rate measured around the yaw axis and the lateral acceleration has been underlined. The analysis has been performed considering an ideal condition for the estimation of the lean angle via inertial sensors. In Section 4.1 the difference between the road and the inertial attitude angles has been underlined. The aim of this Section is to present how the performance of the estimation of the roll angle decrease if some non idealities are introduced. First of all the pitch dynamic is weighed introducing variable speed and slopes on the track. Secondly, banked tracks are considered highlighting the drop off of the performance.

8.2.3.1 Variable speed and slope

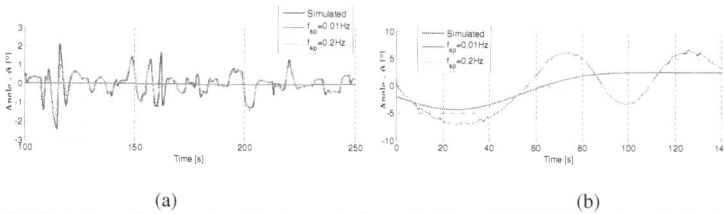

(a)   (b)

Figure 8.14:Pitch dynamics introduced in the track shown in Figure 8.6b (a) Spectral analysis of the pitch dynamic due to the variation of the speed. (b) Spectral analysis of the pitch dynamic due to slopes.

The first considered non ideality is the pitch dynamic that is mainly caused by variable speed and slopes. On the roll estimation point of view, these non ideality can be analyzed at the same time, but it has to be noticed that they have an influence at different bandwidth.

In Figure 8.14 the pitch angle of the motorbike is depicted considering variable speed and slope of the road in a range of $\pm 10\%$ that correspond to an angle α in a range of $\pm 5.7°$. The pitch signal is filtered at different frequencies to highlight the spectral contribution. Comparing Figure 8.14a with Figure 8.14b, it is noticeable that the contribution of the pitch dynamic due to the variable speed is minor than the contribution due to the slope in a bandwidth of 0.2 Hz.

Due to the limited range of the pitch dynamic imposed by the variation of the speed, the drop off of the performance index $J_c\left(\hat{\xi}, f_{sp}, S_I\right)$ due to variable speed shown in Figure 8.15a is close to the performance reduction observed in the simplified condition (Figure 8.13b).

Observing Figure 8.15b, it is noticeable that, if slopes are the main source of the measured pitch dynamic, the most important signal that is needed to perform the estimation of the lean angle is the yaw angular rate while the contribution of the lateral acceleration can be neglected. This conclusion can be explained with the expression of the acquired signals reported in Equation (7.12). The lateral acceleration does not provide any information about the pitch dynamic of the vehicle, then, in a sloped road condition this is not a measure sufficient to perform the estimation of the lean angle with a good performance. It has been highlighted in Section 7.1 that to exactly estimate the lean angle of the two-wheeled vehicle it is necessary to provide also the pitch angle: adopting only the lateral acceleration as input signal to the neural network is not possible to compensate the effect of the pitch dynamic in the lean angle estimation. Therefore, the estimation accuracy reachable with a static identification of the roll angle cannot outperform the accuracy that can be reached with a signal representative of the yaw angular rate.

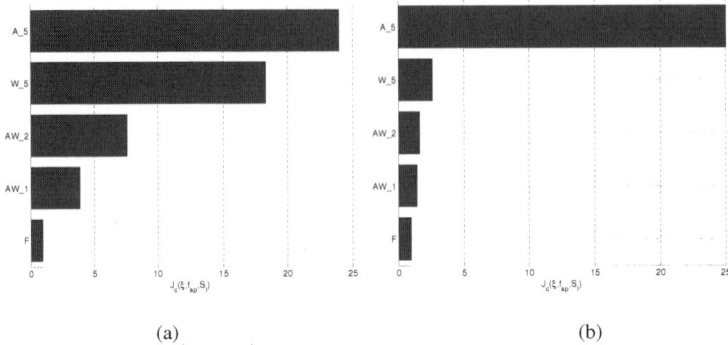

(a) (b)

Figure 8.15: (a) Analysis of $J_c\left(\hat{\xi}, f_{sp}, S_l\right)$ for mixed input configurations with variable speed. (b) Analysis of $J_c\left(\hat{\xi}, f_{sp}, S_l\right)$ for mixed input configurations with sloped road.

In a condition with sloped road, the contribution brings by the measures acquired along the longitudinal axis loose the significance observed in the simplified condition. The interpretation of this result is that adopting the identification scheme of Figure 8.8 the error that is introduced by slopes is more relevant then the correction of the pitch dynamic that can be realized by measuring the longitudinal inertial signals.

8.2.3.2 Bank

The presence of bank of the road is analyzed; tracks considered in the simulation environment are now build with an angle β in the range of $\pm 10°$ (Figure 8.16).

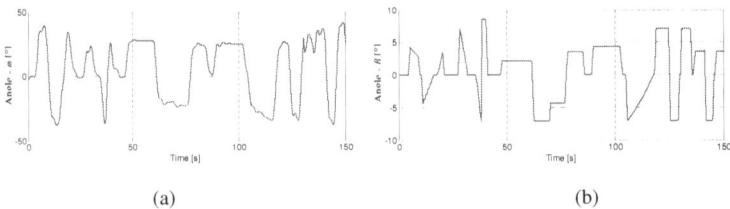

(a) (b)

Figure 8.16. Roll referred to road (a) and bank dynamic (b) introduced in the track shown in Figure 8.6b.

The comparison of $J_c\left(\hat{\xi}, f_{sp}, S_I\right)$ is conducted for mixed configurations considering just the presence of the banked road (Figure 8.17a) and the general condition with all the non ideality that can be introduced in the simulator (Figure 8.17b).

First of all, it is confirmed that the principal measures to perform the lean angle estimation are ω_z and a_y, in fact the performances of the sets F, AW_I and AW_2 are comparable. By the comparison of the two plots, it can be noticed that the loss of performance in a general condition is comparable with the loss of performance due to the simulated bank, therefore the presence of an inclination of the road with an angle β is the main source of error for the estimation of the lean angle of the motorcycle via inertial sensors.

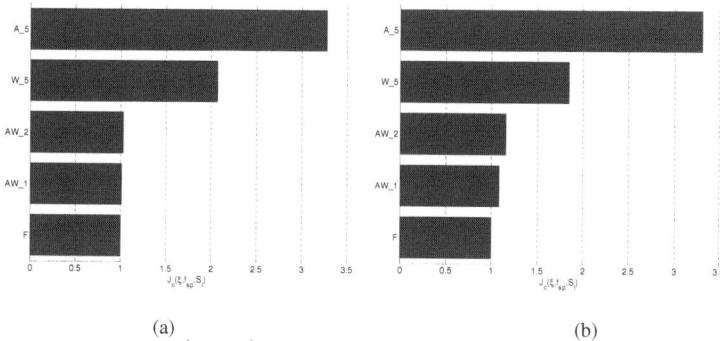

(a)   (b)

Figure 8.17: (a) Analysis of $J_c\left(\hat{\xi}, f_{sp}, S_I\right)$ for mixed input configurations with banked road. (b) Analysis of $J_c\left(\hat{\xi}, f_{sp}, S_I\right)$ for mixed input configurations in general condition.

The previous conclusion is confirmed by the analysis of the bar plot depicted in Figure 8.18 in which the performance index $J_r\left(\hat{\xi}, f_{sp}, F\right)$ is shown for the conditions tested in the simulation environment: the introduction of a bank causes a huge decrease of the estimation accuracy. It has been reported in Section 7.1 that the presence of bank represents an additive error that cannot be compensated via inertial sensors.

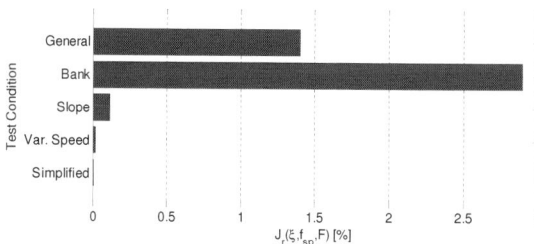

Figure 8.18: Performance index $J_r\left(\hat{\xi}, f_{sp}, F\right)$ for the conditions tested in the simulation environment.

Through the performance analysis based on data provided by the simulator introduced in Section 4.3, it has been underlined that the main source of information to estimate the lean angle of a motorcycle are the lateral acceleration and the yaw angular rate. The measures acquired along the longitudinal axis can be adopted to improve the performance compensating the source of errors that depends on the proper pitch dynamic of the motorbike (dynamic due to the suspension system), while if the measured pitch dynamic is due to presence of slope the significance of these signals is minor and the performance just depend on the acquired yaw rate signal. A bank of the road has been confirmed to be the most relevant cause of decrease of the performance of estimate of the roll angle of a two wheeled vehicle via inertial sensors.

In the next Section, the presented results are validated with the data provided by the experimental test conducted on the circuit of Misano Adriatico.

8.2.4 Performance analysis with experimental data

Performing the frequency analysis on the data provided by the experimental test, the results obtained in the analysis of simulated data are confirmed. The optimum value of the bandwidth of the frequency separation filter is 0.01 Hz if accelerations are considered as input signals and 0.2 Hz if angular rates or mixed configurations are adopted as input signals to the Neural Network. The performance index $J_r\left(\hat{\xi}, f_{sp}, F\right)$ depicted in Figure 8.19 points out that the estimation accuracy reached with a full sensor configuration is mainly due to the contribution provided by the acquired angular rates. Therefore, the performance of estimation via acceleration signals could be improved if they are adopted to identify some components of the angular rates of the motorcycle.

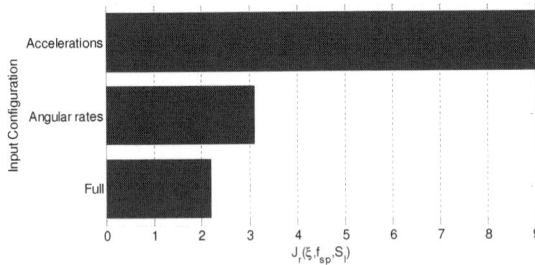

Figure 8.19: Analysis of the performance index $J_r\left(\hat{\xi}, f_{sp}, F\right)$ for different input configurations.

In Figure 8.20 the performance index $J_c\left(\hat{\xi}, f_{sp}, S_I\right)$ is compared as in Section 8.2.2.2 and 8.2.2.3 to analyze the information bring by the acquired inertial signals. It is noticeable that the conclusions underlined by the analysis of the data provided by the simulator are validated.

The bar plots depicted in Figure 8.20a-b remark that the lateral acceleration and the angular rate measured around the yaw axis are the most important measurements to perform the estimation of the roll angle via inertial sensors. Comparing the influence of ω_z and a_y in Figure 8.20c, the yaw angular rate is the foremost signal needed to estimate the roll angle of a motorcycle. A relevant decrease of the performance is due to the elimination of the longitudinal measurement axis, that provides information about the dynamic that are out of the horizontal plane and that are required to compensate the pitch dynamic of the vehicle.

The drop off of the performance depicted in Figure 8.20c is comparable to the decrease of the performance shown in Figure 8.17b: this fact enforce the results obtained with simulated data.

The latest comment regards the small decrease of the performance observed in Figure 8.20a that is due to the low performance that are obtained with the estimation of the lean angle via acceleration signals. The principal cause is that also considering just the low frequency component of the acquired signals, these measurements are affected by a huge amount of noise due to the engine and the translational dynamic of the motorcycle.

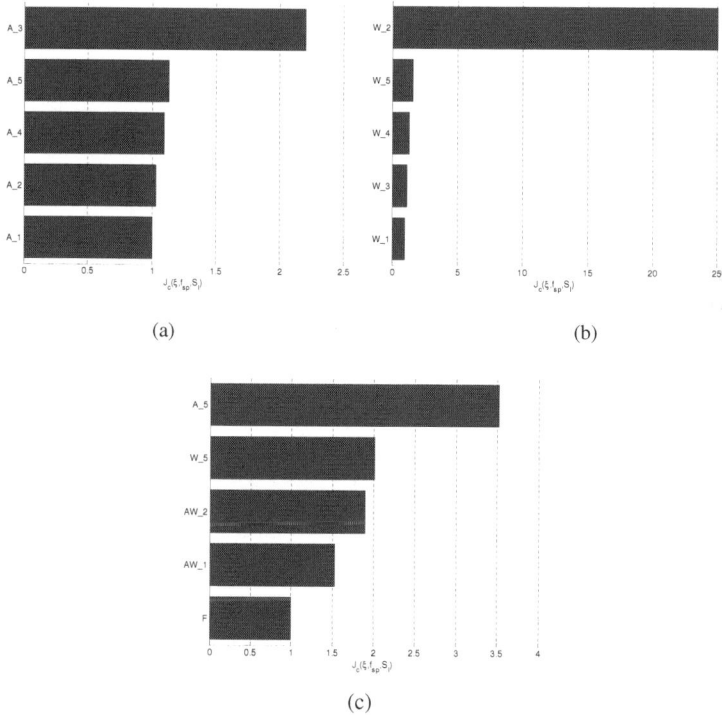

(a)

(b)

(c)

Figure 8.20: Analysis of $J_c\left(\hat{\xi}, f_{sp}, S_I\right)$ with data acquired n the experimental test. (a) Accelerations. (b) Angular rates. (c) Mixed configurations.

In this Section an analysis of the performance of estimation of the roll angle via inertial sensors has been presented. The considered class of algorithms are reported in Figure 8.8 where a frequency split is applied and the low frequency component of the roll angle is estimated with a static non linear function, while the high frequency component is obtained by the integration of the roll angular rate measured on the motorcycle. The results have been supported by an analysis of the model equations of accelerations and angular rates and data provided by a multibody simulator and a test session on the circuit of Santa Monica in Misano Adriatico.

Applying a black-box paradigm on simulated and experimental data it has been shown that the lateral acceleration and the vertical angular rate measured on the motorcycle are strongly related to the low frequency component of the roll dynamic. The information provided by the low frequency component of the measured acquired along the longitudinal axis are to be taken in account to provide relevant information of the non-plane condition of the motorcycle. Moreover the presence of bank of the road is the main source of error that cannot be compensated with inertial sensors.

The foremost conclusion of this Section is that a good performance on the lean angle estimation applying the frequency split principle is reachable just if information about the yaw angular rate is provided by the gyroscope or by the accelerometer. In the following Section this fact will be motivated and used to design model based algorithms for the estimation of the low frequency component of the motorcycle lean angle.

8.3 Estimation of the low frequency component of the roll angle

Herein different algorithms for the estimation of the low frequency component of the lean angle are presented. As depicted in the high level scheme in Figure 7.1, the input data to the estimation algorithm are considered to be unbiased and aligned in the body coordinate system.

The proposed solutions to estimate the low frequency component of the Euler roll angle are based on Equation (8.7) (derived in Chapter 4) that is the lean angle defined by the moments equilibrium to which the vehicle is subject in a steady state turning condition.

$$\varphi_l = \arctan\left(\frac{\dot{\psi}V_x}{g}\right) \tag{8.7}$$

If the expression in Equation (8.7) is adopted to estimate the low frequency component of the roll angle of the vehicle, the problem of estimation of the LF component of the lean angle is reduced to the problem of estimation of the LF component of the yaw rate $\dot{\psi}$.

The main advantages of the roll angle estimation via Equation (8.7) are:

- *Steady state formula*: the expression is just representative of the LF component of the lean angle;
- *Algebraic expression*: the LF component of the signal noise does not cause a drift of the estimation;
- *Performance*: the estimation is based on an angular rate signal and it has been underlined in Section 8.2 that using the angular rates to estimate the LF component of the lean angle, it is possible to obtain an high precision of the estimation (Figure 8.19).

The principal drawbacks of Equation (8.7) have been already discussed in Chapter 4. In particular, considering the approximation $\varphi_E \cong \varphi_l$ and Equation (4.1), the main sources of error between the estimation of φ_l and the reference φ_R are due to:

- *Road bank*: it is not possible to take into account the effect of the angle β;
- *Tires thickness*: the value of $\Delta\varphi_{tyre}$ (Equation (4.6)) cannot be computed;
- *Rider movements*: the guidance style of the rider cannot be modeled and taken into account and it causes an error $\Delta\varphi_{rider}$.

As a consequence, the relation between the estimated lean angle $\hat{\varphi}$ and the reference φ_R can be expressed as

$$\hat{\varphi}_{LF} = \varphi_{I,LF} + \varepsilon_{LF} = \varphi_{R,LF} - \Delta\varphi_{LF,tyre} + \Delta\varphi_{LF,rider} + \beta_{LF} \tag{8.8}$$

where ε is the error of estimation of the inertial lean angle of the vehicle.

In the following the estimation problem of estimation of the yaw rate is approached considering different set of signals:

1. Angular rate $\tilde{\omega}_z$

2. Angular rates $\tilde{\omega}_y$ and $\tilde{\omega}_z$

3. Accelerations $\tilde{a}_y$ and $\tilde{a}_z$ and angular rate $\tilde{\omega}_z$.

The proposed sets of signals underline the importance of the measure $\tilde{\omega}_z$ that result fundamental in all the proposed approach.

First of all the estimation algorithms are separately presented and in Section 8.3.4 they will be compared.

8.3.1 LF component estimation: vertical angular rate

Consider the kinematic contribution of the angular rate measured along the z axis in the body frame:

$$\omega_z = s_\vartheta \dot{\varphi} + c_\varphi c_\vartheta \dot{\psi}. \tag{8.9}$$

In a steady state turning condition $\dot{\varphi} \cong 0$, $\vartheta \cong 0$ and Equation (8.9) reduces to

$$\omega_z \cong c_\varphi \dot{\psi} \tag{8.10}$$

thus, the yaw rate can be estimated as

$$\hat{\dot{\psi}} = \frac{\tilde{\omega}_z}{c_\varphi} \tag{8.11}$$

and substituting in Equation (8.7)

$$t_{\hat{\varphi}} = \frac{V_x \tilde{\omega}_z}{g} \frac{1}{c_{\hat{\varphi}}}, \tag{8.12}$$

then

$$c_\phi t_\phi = s_\phi = \frac{V_x \tilde{\omega}_z}{g}$$

$$\hat{\phi} = \arcsin\left(\frac{V_x \tilde{\omega}_z}{g}\right)$$

$$(8.13)$$

The low frequency component of the lean angle can be estimated with $\tilde{\omega}_z$ as in Figure 8.21. Equation (8.13) has been obtained under the hypothesis that $\varphi_E \cong \varphi_I$. The effect of this assumption can be further analyzed studying the influence of the tires thickness that that can be easily modeled as in Equation (4.6). On the contrary the effect of the rider guidance style cannot be modeled.

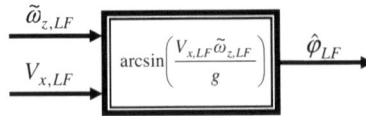

Figure 8.21: Estimation of the LF component of the roll angle through vertical measured angular rate.

The attitude angles that appears in Equation (8.9) are Euler angles, while the lean angle defined by Equation (8.7) is the inertial lean angle, therefore Equation (8.12) should be written as

$$t_{\varphi_I} = \frac{V_x \tilde{\omega}_z}{g} \frac{1}{c_{\varphi_E}}$$

$$(8.14)$$

and recalling Equation (4.5)

$$\cos(\varphi_I + \Delta\varphi)\tan(\varphi_I) = \frac{V_x \tilde{\omega}_z}{g} .$$

$$(8.15)$$

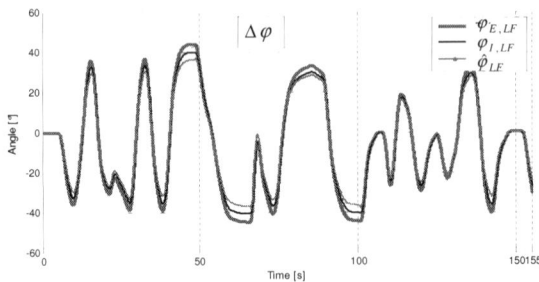

Figure 8.22: Estimation of the LF component of the lean angle with $\tilde{\omega}_z$ - Analysis of the error introduced by the approximation $\varphi_E \cong \varphi_I$.

If the tires thickness is not null, then the argument of the $\arcsin(\cdot)$ function in Equation (8.13) is $\cos(\varphi_I + \Delta\varphi)\tan(\varphi_I)$ and it is easy to verify that

$$\cos(\varphi_I + \Delta\varphi)\tan(\varphi_I) \le \sin(\varphi_I) \quad \forall \Delta\varphi, \varphi_I : \varphi_I \Delta\varphi \ge 0,$$ (8.16)

thus, the algorithm in Figure 8.21 underestimates the inertial roll angle and this error is added to $\Delta\varphi$.

A lap on the track of Figure 8.6b has been simulated with non-constant speed and without banks and slopes and in Figure 8.22 the estimation $\hat{\varphi}_{LF}$ is compared to $\hat{\varphi}_{E,LF}$ and $\hat{\varphi}_{I,LF}$. The proposed algorithm is an angular rate based algorithm, thus, thank to the analysis of Section 8.2, it is posed $f_{sp}=0.2$ Hz. In terms of estimation of the inertial roll angle, an $ESR(\varphi_I) = mse(\varphi_I - \hat{\varphi})/mse(\varphi_I)$ of 0.9% is obtained.

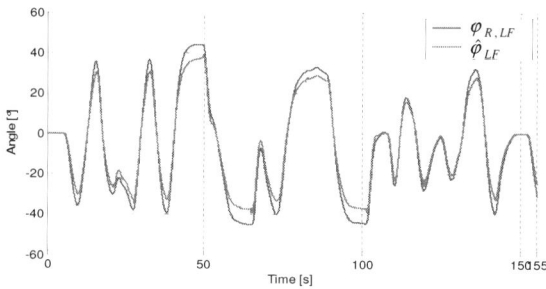

(a)

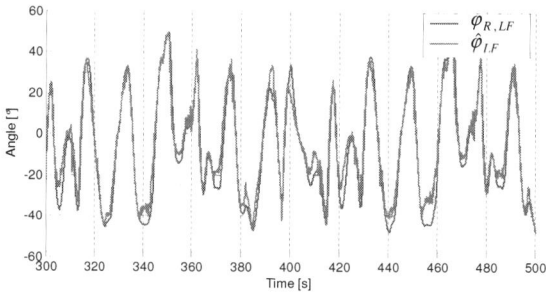

(b)

Figure 8.23: Estimation of the low frequency component of the roll angle applying the algorithm in Figure 8.21 in a simulated environment (a) and in the experimental test (b).

In Figure 8.23 estimation results are depicted in a simulated environment (single lap of circuit in Figure 8.6b covered in a general condition) and experimental environment respectively (test on Enzo Ferrari circuit).

Again, the estimation performance is described with the ESR of the estimation error that is

$$J_{LF}\left(f_{sp}\right) = \frac{mse\left(\varphi_{R,LF,f_{sp}} - \hat{\varphi}_{LF,f_{sp}}\right)}{mse\left(\varphi_{LF,f_{sp}}\right)}. \tag{8.17}$$

The performances for the algorithm presented in this Section are $J_{LF}(0.2) \cong 2.7\%$ for the simulated data and $J_{LF}(0.2) \cong 5.7\%$ for data acquired during the experimental test: the descerase of the performance is caused by noise condition in which the experimental data are acquired. Notice that in Figure 8.23b the effect of the approximation $\varphi_E \cong \varphi_I$ is not as evident as in Figure 8.23a: this is due to the effect of the rider movements that is typically in accordance with the displacement of the tyres POC caused by the tyres thickness.

8.3.2 LF component estimation: lateral and vertical angular rates
Three algorithms based on the information provided by the lateral and vertical acquired angular rates are herein proposed.

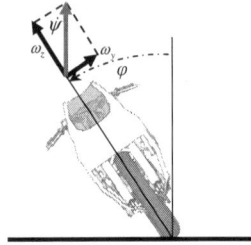

Figure 8.24: Rappresentation of the contribution of the vehicle yaw rate on the measured vertical (ω_z) and lateral (ω_y) angular rates.

As described by the model in Equation (8.6) and depicted in Figure 8.24 both the vertical and lateral measured angula rates are informative of the yaw rate of the vehicle.

The first algorithm estimates the yaw rate of the vehicle starting from the information provided by $\tilde{\omega}_y$ (as it has been done in the previous Section with $\tilde{\omega}_z$). The second proposed algorithm is based on the estimation of $\dot{\psi}$ considering the information provided both by $\tilde{\omega}_y$ and $\tilde{\omega}_z$. The last algorithm is based on the combination of the output of the algorithm presented in Section 8.3.1 and 8.3.2.1.

8.3.2.1 Lateral angular rate based algorithm
The angular rate measured along the y body axis, can be described as

$$\omega_y = \dot{\vartheta} + s_\varphi \dot{\psi} \tag{8.18}$$

and in a steady state turning condition it can be approximated by

$$\omega_y \cong s_\varphi \dot{\psi}.$$ (8.19)

In Section 7.3, it has been observed that $\tilde{\omega}_y \leq 0$ both for right and left turn, thus, from Equation (8.19) the yaw rate module can be estimated as

$$\left|\dot{\hat{\psi}}\right| = \left|\frac{\tilde{\omega}_y}{s_\varphi}\right|.$$ (8.20)

Considering $\varphi_E \cong \varphi_I$ and substituting Equation (8.20) in Equation (8.7)

$$t_{|\phi|} = \frac{V_x \left|\tilde{\omega}_y\right|}{g} \frac{1}{s_{|\phi|}}.$$ (8.21)

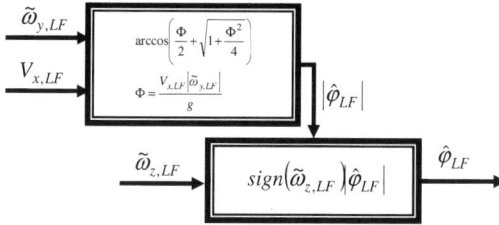

Figure 8.25: Estimation of the LF component of the roll angle through lateral measured angular rate.

The previous Equation can be further elaborated (for the sake of conciseness the module operator end the symbol tilde are omitted):

$$
\begin{aligned}
t_\varphi &= \frac{V_x \omega_y}{g} \frac{1}{s_\varphi} \\
s_\varphi^2 &= \frac{V_x \omega_y}{g} c_\varphi \\
c_\varphi^2 &+ \frac{V_x \omega_y}{g} c_\varphi - 1 = 0
\end{aligned}
$$ (8.22)

thus,

$$
\begin{aligned}
\left|\hat{\varphi}\right| &= \arccos\left(\frac{\Phi}{2} + \sqrt{1 + \frac{\Phi^2}{4}}\right) \\
\Phi &= \frac{V_x \left|\tilde{\omega}_y\right|}{g}
\end{aligned}
$$ (8.23)

The sign of the estimated roll angle can be recovered with the sign of the angular rate measured along the z body axis as

$$\hat{\varphi} = -\text{sign}\,(\tilde{\omega}_z)\arccos\left(\frac{\Phi}{2} + \sqrt{1 + \frac{\Phi^2}{4}}\right). \tag{8.24}$$

The algorithm to estimate the module of the LF component of the roll angle through $\tilde{\omega}_y$ is depicted in Figure 8.25.

As it has been done in the previous Section, the error introduced in Equation (8.21) by the approximation $\varphi_E \cong \varphi_I$ can be studied. Ignoring the module operator and explicating the inertial and Euler roll angle, Equation (8.21) can be reworked as

$$t_{\varphi_I} = \frac{V_x \tilde{\omega}_y}{g} \frac{1}{s_{\varphi_E}} \tag{8.25}$$

thus

$$\sin(\varphi_I + \Delta\varphi)\tan(\varphi_I) = \frac{V_x \tilde{\omega}_y}{g}. \tag{8.26}$$

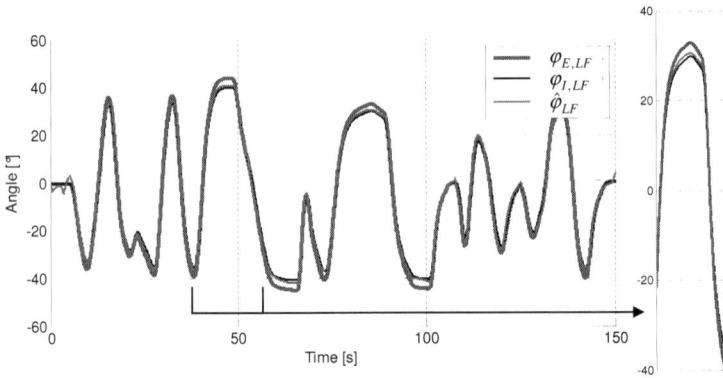

Figure 8.26: Estimation of the LF component of the lean angle with $\tilde{\omega}_y$ - Analysis of the error introduced by the approximation $\varphi_E \cong \varphi_I$.

The argument of the $\arccos(\cdot)$ in Equation (8.23) can be studied if the tires thickness is greater than zero and it is easy to show that

$$\frac{\sin(\varphi_I +\Delta\varphi)\tan(\varphi_I)}{2}+\sqrt{1+\left(\frac{\sin(\varphi_I +\Delta\varphi)\tan(\varphi_I)}{2}\right)^2} \geq$$

$$\frac{\sin(\varphi_I)\tan(\varphi_I)}{2}+\sqrt{1+\left(\frac{\sin(\varphi_I)\tan(\varphi_I)}{2}\right)^2} \quad , \tag{8.27}$$

$$\Delta\varphi, \varphi_I : \varphi_I \Delta\varphi \geq 0$$

in fact condition (8.27) reduces to

$$[\sin(\varphi_I)-\sin(\varphi_I +\Delta\varphi)]^2 \geq 0 \quad \forall\, \Delta\varphi, \varphi_I : \varphi_I \Delta\varphi \geq 0, \tag{8.28}$$

thus, due to the tires thickness the estimation algorithm depicted in Figure 8.25 overestimate the inertial roll angle as depicted in Figure 8.26 in which the simulated data on circuit Figure 8.6b with variable speed are considered. Considering $f_{sp}=0.2$ Hz, an $ESR(\varphi_I)$ of 0.4% is obtained.

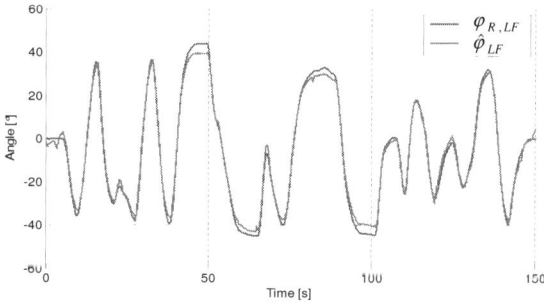

(a)

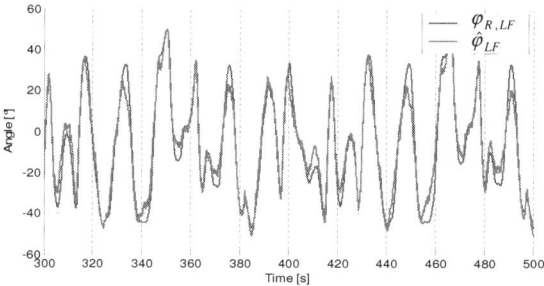

(b)

Figure 8.27: Estimation of the low frequency component of the roll angle applying the algorithm in Figure 8.25 and recovering the sign as in Equation (8.24) in a simulated environment (a) and in the experimental test (b).

In Figure 8.27 estimation results are depicted in a simulated environment (single lap of circuit in Figure 8.6b covered in real condition) and experimental environment respectively (test on Enzo Ferrari circuit).

The performances for the algorithm presented in this Section are $J_{LF}(0.2) \cong 0.8\%$ for the simulated data and $J_{LF}(0.2) \cong 5.7\%$ for data acquired during the experimental test in Imola.

8.3.2.2 Yaw rate estimation with lateral and vertical angular rates

The information provided by the signals $\tilde{\omega}_z$ and $\tilde{\omega}_y$ can also be combined to estimate the yaw rate $\dot{\psi}$ of the vehicle.

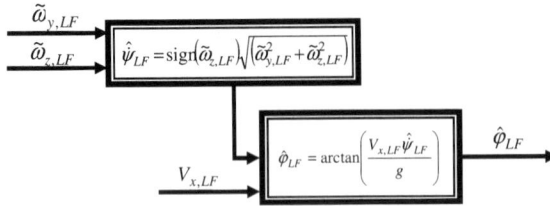

Figure 8.28: Estimation of the LF component of the roll angle through lateral and vertical measured angular rates.

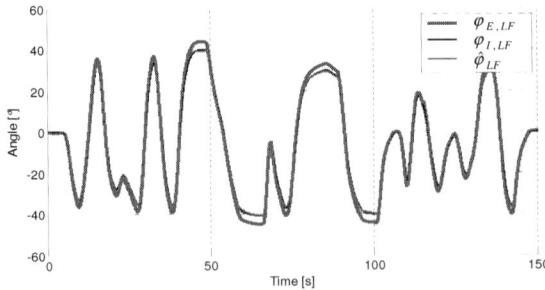

Figure 8.29: Estimation of the LF component of the lean angle with $\tilde{\omega}_y$ and $\tilde{\omega}_z$ - Analysis of the error introduced by the approximation $\varphi_E \cong \varphi_I$.

Consider now Equation (8.10) and (8.18): in a steady state turning condition, the yaw rate $\hat{\dot{\psi}}$ of the vehicle can be estimated as

$$\hat{\dot{\psi}} = \text{sign}(\tilde{\omega}_z)\sqrt{(\tilde{\omega}_y^2 + \tilde{\omega}_z^2)} \qquad (8.29)$$

and the lean angle can be estimated as

$$\hat{\varphi} = \arctan\left(\frac{V_x \hat{\dot{\psi}}}{g}\right). \qquad (8.30)$$

In Figure 8.28 it is depicted the scheme for the estimation of the LF component of the roll angle. Again it is considered that $\varphi_E \cong \varphi_I$, but in this formulation other terms of error than $\Delta\varphi$ are not introduced and the LF component of the inertial roll angle is recovered as shown in Figure 8.29, in which the simulated data on circuit Figure 8.6b with variable speed are considered.

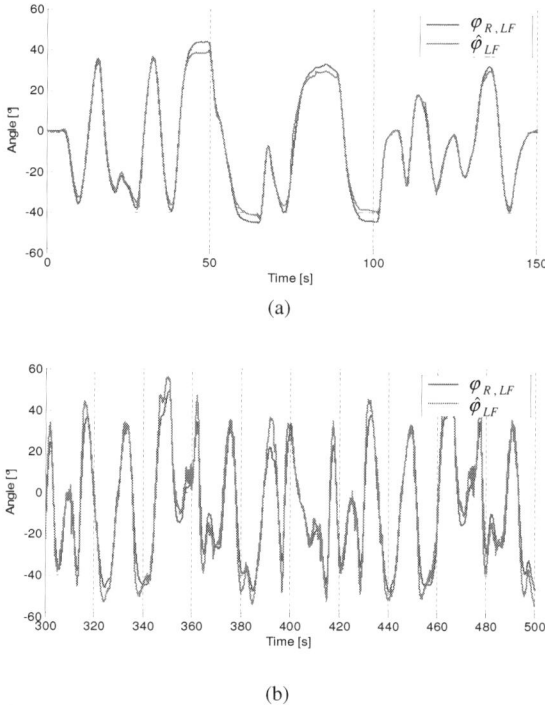

(a)

(b)

Figure 8.30: Estimation of the low frequency component of the roll angle applying the algorithm in Figure 8.28 in a simulated environment (a) and in the experimental test (b).

In Figure 8.30 the estimation results are shown considering the same simulated and experimental conditions of Figure 8.23 and Figure 8.27.

The performances for the algorithm presented in this Section are $J_{LF}(0.2) \cong 0.8\%$ for the simulated data and $J_{LF}(0.2) \cong 6\%$ for data acquired during the experimental test in Imola.

8.3.2.3 Linear combination of lateral and vertical estimation

The angular rates measured along the y and z axis of the body coordinate system, can also be combined observing that in a steady state turning condition

$$\begin{cases} \tilde{\omega}_y - \dot{\psi} \geq \tilde{\omega}_z - \dot{\psi}, & \varphi \leq 45° \\ \tilde{\omega}_y - \dot{\psi} < \tilde{\omega}_z - \dot{\psi}, & \varphi > 45° \end{cases} \tag{8.31}$$

that means that the estimation of the LF component of the lean angle realized with $\tilde{\omega}_z$ gives better performance if $\varphi \leq 45°$, while the LF component of the lean angle estimated by $\tilde{\omega}_y$ gives better performance if $\varphi > 45°$. Thank to this remark the estimation performed by the algorithms in Figure 8.21 and Figure 8.25 can be combined as a function of the output estimation.

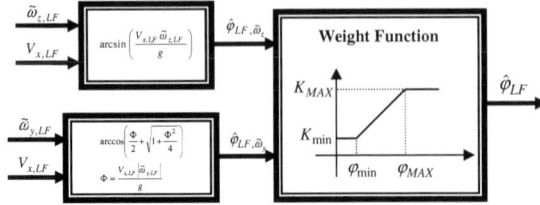

Figure 8.31: Estimation of $\hat{\varphi}_{LF}$ combining $\hat{\varphi}_{LF,\tilde{\omega}_y}$ and $\hat{\varphi}_{LF,\tilde{\omega}_z}$.

Now consider the weighted mean

$$\hat{\varphi}_{LF} = \left(1 - K\left(\hat{\varphi}_{LF}\right)\right)\hat{\varphi}_{LF,\tilde{\omega}_z} + K\left(\hat{\varphi}_{LF}\right)\hat{\varphi}_{LF,\tilde{\omega}_y} \qquad (8.32)$$

where $\hat{\varphi}_{LF,\tilde{\omega}_z}$ and $\hat{\varphi}_{LF,\tilde{\omega}_z}$ are the estimation achieved with $\tilde{\omega}_z$ and $\tilde{\omega}_y$ respectively. The weight function $K\left(\hat{\varphi}_{LF}\right)$ can be defined as shown in Figure 8.31:

$$K\left(\hat{\varphi}_{LF}\right) = \begin{cases} K_{\min} & \hat{\varphi}_{LF} \leq \varphi_{\min} \\ m_{\hat{\varphi}}\hat{\varphi}_{LF} + q_{\hat{\varphi}} & \varphi_{\min} < \hat{\varphi}_{LF} < \varphi_{MAX} \\ K_{MAX} & \hat{\varphi}_{LF} \geq \varphi_{MAX} \end{cases} \qquad (8.33)$$

where

$$m_{\hat{\varphi}} = \frac{K_{MAX} - K_{\min}}{\varphi_{MAX} - \varphi_{\min}}, \qquad (8.34)$$

thus, substituting Equation (8.33) in Equation (8.32),

$$\hat{\varphi}_{LF} = \begin{cases} \left(1 - K_{\min}\right)\hat{\varphi}_{LF,\tilde{\omega}_z} + K_{\min}\hat{\varphi}_{LF,\tilde{\omega}_y} & if \quad \hat{\varphi}_{LF} \leq \varphi_{\min} \\ \dfrac{\hat{\varphi}_{LF,\tilde{\omega}_z} + q_{\hat{\varphi}}\left(\hat{\varphi}_{LF,\tilde{\omega}_y} - \hat{\varphi}_{LF,\tilde{\omega}_z}\right)}{1 + m_{\hat{\varphi}}\left(\hat{\varphi}_{LF,\tilde{\omega}_z} + \hat{\varphi}_{LF,\tilde{\omega}_y}\right)} & if \quad \varphi_{\min} < \hat{\varphi}_{LF} < \varphi_{MAX}. \\ \left(1 - K_{MAX}\right)\hat{\varphi}_{LF,\tilde{\omega}_z} + K_{MAX}\hat{\varphi}_{LF,\tilde{\omega}_y} & if \quad \hat{\varphi}_{LF} \geq \varphi_{MAX} \end{cases} \qquad (8.35)$$

The parameters K_{min}, K_{MAX}, φ_{min} and φ_{MAX} can be chosen so that the cost function $J_{LF}(f_{sp})$ is minimized. Here the chosen values for the parameters of the weighting function are: $K_{min} = 0.2$, $K_{MAX} = 0.8$, $\varphi_{min} = 15°$ and $\varphi_{MAX} = 45°$.

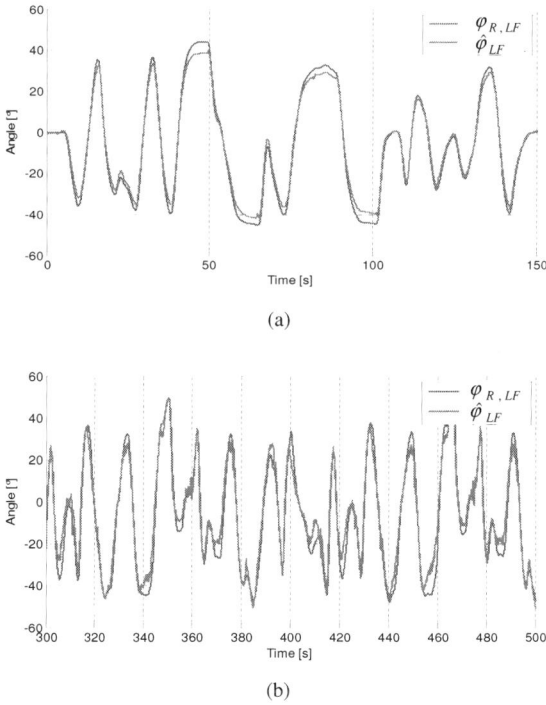

(a)

(b)

Figure 8.32: Estimation of the low frequency component of the roll angle applying Equation (8.35) in a simulated environment (a) and in the experimental test (b).

In Figure 8.32 the estimation results are shown in the simulated and experimental condition of Figure 8.30. The performances for the algorithm presented in this Section are $J_{LF}(0.2) \cong 1.3\%$ for the simulated data and $J_{LF}(0.2) \cong 4.7\%$ for data acquired during the experimental test in Imola.

8.3.3 LF component estimation: lateral and vertical accelerations

The yaw rate of the vehicle can be estimated also considering the acquired inertial accelerations. In steady state turning condition, Equation (7.12) reduces to

$$a_y \cong s_\varphi g + c_\varphi \dot{\psi} N_x$$
$$a_z \cong c_\varphi g - s_\varphi \dot{\psi} N_x$$

$$(8.36)$$

it follows that

$$a_y^2 + a_z^2 \cong g^2 + \dot{\psi}^2 V_x^2 \tag{8.37}$$

and

$$\left|\hat{\dot{\psi}}\right| = \sqrt{\frac{\left(\tilde{a}_y^2 + \tilde{a}_z^2\right) - g^2}{V_x^2}} \,. \tag{8.38}$$

The sign of the estimated yaw rate can be computed using the sign of the angular rate measured along the z axis of the body coordinate system:

$$\hat{\dot{\psi}} = \text{sign}(\tilde{\omega}_z) \sqrt{\frac{\left(\tilde{a}_y^2 + \tilde{a}_z^2\right) - g^2}{V_x^2}} \,. \tag{8.39}$$

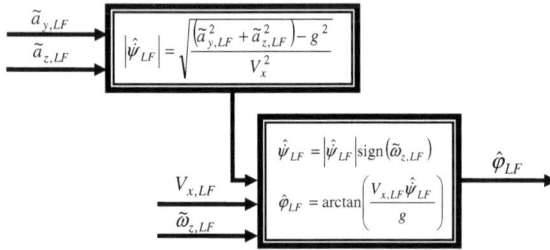

Figure 8.33: Estimation of the LF component of the roll angle through lateral and vertical measured accelerations.

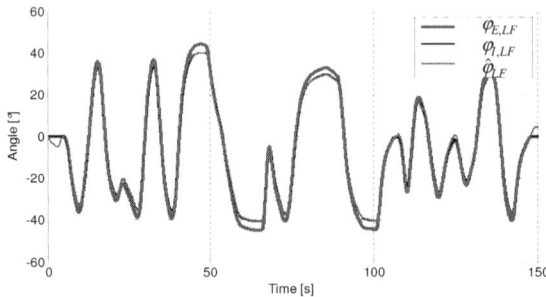

Figure 8.34: Estimation of the LF component of the lean angle with $\tilde{a}_y$ and $\tilde{a}_z$ - Analysis of the error introduced by the approximation $\varphi_E \cong \varphi_I$.

As a result, the lean angle of the vehicle can be estimated as in Equation (8.30). In Figure 8.33, the estimation algorithm to obtain the LF component of the lean angle is shown.

The estimated $\hat{\psi}$ is not affected by the approximation of the Euler roll angle with the inertial one, then, as shown in Figure 8.34 in which the circuit depicted in Figure 8.6b is covered with variable speed, the difference between the estimated $\hat{\varphi}_{LF}$ and the reference φ_E is mainly due to $\Delta\varphi$.

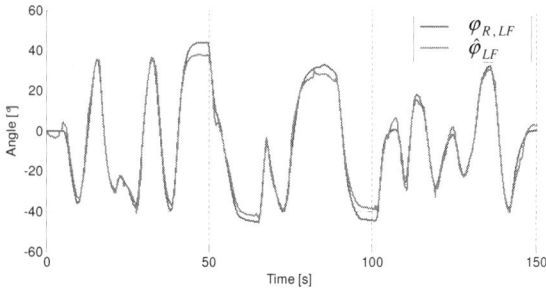

(a)

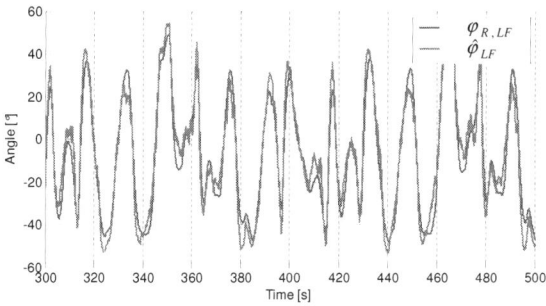

(b)

Figure 8.35: Estimation of the low frequency component of the roll angle applying the scheme depicted in Figure 8.33 in a simulated environment (a) and in the experimental test (b).

The algorithm proposed in Figure 8.33 is validated in simulation (single lap of circuit in Figure 8.6b covered in real condition) and with experimental data (test on Enzo Ferrari circuit). The estimation results are shown in Figure 8.35. The performances are $J_{LF}(0.2) = 0.9\%$ for the simulated data and $J_{LF}(0.2) = 7.7\%$ for data acquired during the experimental test in Imola.

8.3.4 Performance comparison
The performances of the proposed algorithms for the estimation of the low frequency component of the lean angle are now compared.

In Table 8.1, different conditions have been taken into account and for each of them the mean of the performances over various data sets is reported (*i.e.* all the runs of the experimental sets of data are considered and the average performance is computed):

- *Simplified condition*: data provided by the simulator using the circuits depicted in Figure 8.6 covered at constant speed and without slopes and banks. In this condition the difference between the Euler and inertial roll angle is the major cause of error;
- *General condition*: data provided by the simulator using the circuits depicted in Figure 8.6 covered with variable speed and simulated slopes and banks. In this condition the validity of the Equation (8.5) is evaluated to highlight the effect of the motorcycle pitch dynamic and bank of the road;
- *Experimental test*: data collected during the tests in Misano and Imola. This the real condition in which the estimation has to be performed, thus, not only the effect of the motorcycle dynamic are evaluated, but also the error induced by the non ideal mounting position of the inertial platform that is not fixed in COG of the vehicle and the signal noise.

Estimation Algorithm	Analyzed Condition					
	Simplified Simulation		*General Simulation*		*Experimental Test*	
Figure 8.21	$J_{LF}(0.2)$	1.7%	$J_{LF}(0.2)$	3.9%	$J_{LF}(0.2)$	7.5%
	J_{φ_R}	1.8%	J_{φ_R}	4.2%	J_{φ_R}	8.0%
Figure 8.25	$J_{LF}(0.2)$	0.6%	$J_{LF}(0.2)$	3.25%	$J_{LF}(0.2)$	4.9%
	J_{φ_R}	0.7%	J_{φ_R}	3.8%	J_{φ_R}	5.8%
Figure 8.28	$J_{LF}(0.2)$	0.7%	$J_{LF}(0.2)$	3.5%	$J_{LF}(0.2)$	4.8%
	J_{φ_R}	0.7%	J_{φ_R}	4.0%	J_{φ_R}	5.0%
Figure 8.31	$J_{LF}(0.2)$	0.8%	$J_{LF}(0.2)$	3.2%	$J_{LF}(0.2)$	5.0%
	J_{φ_R}	0.9%	J_{φ_R}	3.6%	J_{φ_R}	6.5%
Figure 8.33	$J_{LF}(0.2)$	0.6%	$J_{LF}(0.2)$	5.45%	$J_{LF}(0.2)$	7.0%
	J_{φ_R}	0.7%	J_{φ_R}	5.8%	J_{φ_R}	7.5%

Table 8.1: Comparison performance of estimation of the road lean angle of a motorcycle with frequency serration algorithm.

The performance indices summarized in Table 8.1 are $J_{LF}(f_{sp})$ to describe the estimation of the LF component of the roll angle $\varphi_{R,LF}$, and J_{φ_R} (Equation (8.40)) to characterize the performance of the estimation of the entire roll angle φ_R applying the frequency separation principle. The comparison between $J_{LF}(f_{sp})$ and J_{φ_R} shows up that the performance of the frequency separation based algorithms mainly depend on the estimation of the LF component of the roll angle, thus in what follows the limit of the LF estimation algorithm will be pointed out.

$$J_{\varphi_R} = \frac{mse(\varphi_R - \hat{\varphi})}{mse(\varphi_R)} \tag{8.40}$$

8.3.4.1 Simulation results: simplified condition

In the simplified condition the worst performance are reached by the algorithm based on the angular rate $\tilde{\omega}_z$: this result is motivated by the effect of the approximation $\varphi_E \cong \varphi_I$. A sensitivity analysis of the effect of the tires thickness $\Delta\varphi$ is useful for the explanation of the result.

In Section 8.3.1 and 8.3.2, it has been shown that due to the tires thickness and the approximation $\varphi_E \cong \varphi_I$, the estimation of $\hat{\varphi}$ performed with $\tilde{\omega}_z$ and $\tilde{\omega}_y$ can be expressed as

$$\begin{aligned}
\hat{\varphi}_{\tilde{\omega}_y} &= \arccos\left(A_{\tilde{\omega}_y}\right) \\
\hat{\varphi}_{\tilde{\omega}_z} &= \arcsin\left(A_{\tilde{\omega}_z}\right)
\end{aligned} \tag{8.41}$$

where

$$\begin{aligned}
A_{\tilde{\omega}_y} &= \frac{\sin(\varphi_I + \Delta\varphi)\tan(\varphi_I)}{2} + \sqrt{1 + \left(\frac{\sin(\varphi_I + \Delta\varphi)\tan(\varphi_I)}{2}\right)^2} \ . \\
A_{\tilde{\omega}_z} &= \cos(\varphi_I + \Delta\varphi)\tan(\varphi_I)
\end{aligned} \tag{8.42}$$

Now consider $S_{\tilde{\omega}_y}$ and $S_{\tilde{\omega}_z}$ defined in Equation (8.43), that represent the sensitivity to $\Delta\varphi$ of the arguments $A_{\tilde{\omega}_y}$ and $A_{\tilde{\omega}_z}$.

$$\begin{aligned}
S_{\tilde{\omega}_y} &= \frac{\partial A_{\tilde{\omega}_y}}{\partial \Delta\varphi} \\
S_{\tilde{\omega}_z} &= \frac{\partial A_{\tilde{\omega}_z}}{\partial \Delta\varphi}
\end{aligned} \tag{8.43}$$

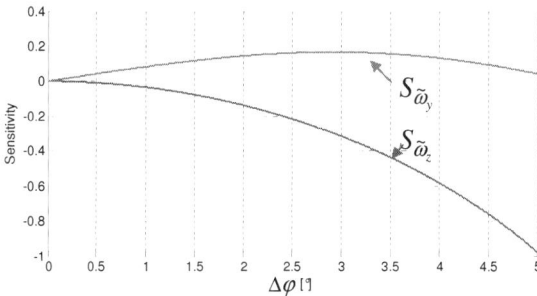

Figure 8.36: Sensitivity analysis to the error introduced in Equation and by $\Delta\varphi$.

In Figure 8.36 $S_{\tilde{\omega}_y}$ and $S_{\tilde{\omega}_z}$ are depicted as a function of $\Delta\varphi$. Recalling Section 8.3 and considering $\varphi_l = 10\Delta\varphi$ (without loss of generality $\Delta\varphi \geq 0$ is considered), it can be remarked that:

- $S_{\tilde{\omega}_y}(\Delta\varphi) \geq 0$ and $S_{\tilde{\omega}_z}(\Delta\varphi) \leq 0$, thus it is confirmed that Equation (8.23) overestimate φ_l, while Equation (8.13) underestimate φ_l;

- $\left|S_{\tilde{\omega}_z}\right| >> \left|S_{\tilde{\omega}_y}\right|$, thus the estimation of the LF component of the roll angle via $\tilde{\omega}_z$ has an higher sensitivity to $\Delta\varphi$.

On one hand, it should be now clear that the loss of performance of estimation algorithm based on $\tilde{\omega}_z$ is due to the high sensitivity of $A_{\tilde{\omega}_z}$ to $\Delta\varphi$; on the other hand, thanks to the low sensitivity of $A_{\tilde{\omega}_y}$ to $\Delta\varphi$, the performance of the estimation algorithm based on $\tilde{\omega}_y$ is comparable to the performance of the estimation scheme in Figure 8.28, Figure 8.31 and Figure 8.33.

8.3.4.2 Simulation results: general condition

It is again noticeable that the algorithm based on $\tilde{\omega}_z$ has worse performance with respect to the other angular rates based algorithms.

The algorithm based on lateral and vertical accelerations is characterized by the greatest reduction of the performance: this result is explained by the fact that in a real operating condition the translational contributions of the motorcycle dynamic strongly affect the estimation of the yaw rate of the vehicle.

Estimation Algorithm	Linear correlation				
	$\left|\rho_\vartheta\right|$	$\left|\rho_{\dot\vartheta}\right|$	$\left|\rho_\varphi\right|$	$\left|\rho_{\dot\varphi}\right|$	$\left|\rho_\beta\right|$
Figure 8.21	0.12	0.12	0.51	0.30	0.52
Figure 8.25	0.10	0.16	0.37	0.27	0.77
Figure 8.28	0.15	0.13	0.36	0.31	0.77
Figure 8.31	0.10	0.14	0.31	0.32	0.68
Figure 8.33	0.14	0.14	0.27	0.24	0.63

Table 8.2: Linear correlation between estimation error and sources of error for frequency separation based algorithms.

The correlation between the estimation error and the principal sources of performance degradation is studied. In Table 8.2, for each of the proposed algorithms, the absolute value of the linear correlation coefficient ρ and:

- Pitch ($\left|\rho_\vartheta\right|$)

- Pitch rate ($\left|\rho_{\dot\vartheta}\right|$)

- Roll angle ($\left|\rho_\varphi\right|$)

- Roll rate ($\left|\rho_{\dot\varphi}\right|$)

- Bank ($\left|\rho_\beta\right|$)

is detailed. The reported values of the linear correlation are the mean of the values computed simulating the circuits in Figure 8.6 in a realistic condition.

Observing the results in Table 8.2, it can be concluded that:

- The bank of the road is the most relevant source of error. It has been already underlined by the performance analysis with Neural Network that the principal term of error that affect the estimation performance is the presence of banks on the road that cannot be identified with inertial signals: the presence of banked road on the simulated track cause a relevant drop of the performance of all the proposed algorithms with respect to the simplified case.

- The estimation error is strongly related to the inertial roll angle. The error $\Delta\varphi$ caused by the assumption $\varphi_E \cong \varphi_I$ is proportional to the roll angle of the motorcycle and it has been underlined in Section 8.3.4.1 that the algorithm in Figure 8.21 has the highest sensitivity to this error;

- All the proposed algorithms are based on the hypothesis of steady state turning condition, as a consequence the correlation of the estimation error with the roll rate is relevant;

- The error is also partially correlated to the pitch dynamic of the vehicle. By the definition of the rotation matrix, it has been highlighted that the knowledge of the pitch angle is necessary to correctly estimate the lean angle of the motorcycle; all proposed algorithms do not estimate the pitch Euler angle of the motorbike.

8.3.4.3 Experimental results

The experimental results confirm what has been underlined with the simulation analysis.

Due to the effect of the translational accelerations that are not considered in model (8.36), the algorithm based on the lateral and vertical accelerations has the worst performance.

Among the algorithms based on angular rates measurement, the scheme in Figure 8.21 seems to be the one with worse capability of fitting the lean angle of the vehicle.

A last important remark regards the effect of the mounting attitude on the performance estimation. Let $\mathbf{A}$ be a kinematic vector described in the inertial coordinate system, $\mathbf{a}_o$ the same vector described in a frame with a fixed rotation $(\varphi_o, \vartheta_o, \psi_o)$ with respect to the body frame. The vectors $\mathbf{A}$ and $\mathbf{a}_o$ are related by

$$\mathbf{a}_o = R_{ZXY}(\varphi_o, \vartheta_o, \psi_o) R_{ZXY}(\varphi, \vartheta, \psi) \mathbf{A} . \tag{8.44}$$

If $(\Delta\varphi_o, \Delta\vartheta_o, \Delta\psi_o)$ is the error of estimation of $(\varphi_o, \vartheta_o, \psi_o)$, the vector $\mathbf{a}$ in the body coordinate system is defined by

$$\mathbf{a} = R_{ZXY}(\Delta\varphi_o, \Delta\vartheta_o, \Delta\psi_o) R_{ZXY}(\varphi, \vartheta, \psi) \mathbf{A} \tag{8.45}$$

and

$$\mathbf{a} = R_{ZXY}\big(\widetilde{\varphi}, \widetilde{\vartheta}, \widetilde{\psi}\big)\mathbf{A} \neq R_{ZXY}\big(\Delta\varphi_o + \varphi, \Delta\vartheta_o + \vartheta, \Delta\psi_0 + \psi\big)\mathbf{A}, \tag{8.46}$$

thus, if residual errors due mounting attitude estimation persist, the acquired inertial signals would estimate $\dot{\widetilde{\psi}} \neq \dot{\psi}$ and $\widetilde{\varphi} \neq \varphi$ and just for small values of the triple $\big(\Delta\varphi_o, \Delta\vartheta_o, \Delta\psi_o\big)$ can be guaranteed that $\widetilde{\varphi} \cong \varphi$.

8.4 Concluding remarks and future works

In this Chapter the frequency separation principle has been presented and applied to the estimation of the roll angle of a motorcycle: the problem of estimation of the lean angle of the motorcycle is tackled considering a linear combination of a high frequency and a low frequency observer.

On one hand, the roll gyroscope is affected by a low frequency noise and just the high frequency component of the lean angle estimation can be recovered by the integration of ω_x. On the other hand, the low frequency component of the acquired inertial signals (aligned in the body frame and bias compensated) do not suffer of drift error and they can be adopted to estimate the low frequency component of the lean angle. The estimation of φ_{LF} is not trivial. In this Chapter an analysis in the Neural Network framework is firstly proposed recovering knowledge about the signals that have to be employed to optimize the estimation performance. This information have been adopted to develop model based algorithms that guarantee an estimation performance of 6% (ESR). The estimation accuracy can be improved reducing the influence of the pitch dynamic in the roll estimation. Pitch angle estimation algorithms based on the frequency separation principle are currently under studying. In a Kalman Filtering framework both the roll angle and pitch angle can be estimated.

Chapter 9
Motorcycle attitude estimation with inertial sensors: Kalman filtering

In Chapter 3, the Kalman Filter has been presented. In this Chapter, the theory of Kalman filtering is applied to solve the problem of attitude estimation of a motorcycle. The direct formulation of the Kalman Filter for attitude estimation is adopted (direct estimation of the attitude) proposing a new model for the estimation with inertial sensors [31], [96]. The EKF and the UKF are compared underlining the main advantages of the unscented transformation. As for the algorithms proposed in Chapter 8, the input data are considered to be aligned in the body frame and unbiased. The analysis is completed with a comparison between the attitude estimation performed with an EKF and without the errors signals representation and the observation performed by an EKF with signals errors estimation.

In a model-based estimation problem such as the Kalman filter, the filtering (estimation) performance is inevitably much dependent upon how well the physics arising in an actual system are reflected in the stochastic plant model. It is natural that more accurate estimates can be expected as the model gets closer to the real. Thus, the Chapter is organized as follows. First of all the model of the system to be identify is discussed (Section 9.1). Different approaches are described and for each of them advantages and drawbacks will be underlined. It will be highlighted that the principal constrains are due to the available quantities to be measured and to the computational capability of the ECU that equips the motorbike. Then the Extended Kalman Filter is introduced and the discrete implementation is presented (Section 9.2). The Unscented Kalman Filter is also applied to the attitude estimation problem (Section 9.3), Both EKF and UKF are tuned (Section 9.4) and the performance are compared (Section 9.5). In Section 9.6, the performances of the Extended Kalman Filter are compared to the performances of the Frequency Separation algorithms to underline the robustness of the Kalman filtering approach to the pitch dynamic of the vehicle.

Finally, some concluding remarks and future works are presented (Section 9.6).

9.1 Model definition

The problem of definition of the proper model in the attitude estimation context is not trivial. Here, the Euler angles parameterization is used even if it brings to a non-linear description of the process. The adopted model strictly depends of the available signals on the vehicle. In avionic applications (*e.g.* [34], [13], [97]), the vehicle dynamic can be fully measured and described combining positioning signals, angular rate signals, fixed star measurements and magnetic field measurements. In robotics just angular rates and accelerations are available and the process can be modeled considering a switching state observer [98]. In automotive applications different sets of sensors are used just to describe a part of the vehicle dynamic. The signals provided by a IMU are typically employed in different applications (see [99]): yaw control [100], roll stability control, steering control [54].

In Section 7.1, the models of the acquired signals have been presented. In particular the model of the measured accelerations has been widely discussed. Here, the representations of the measured angular rates and accelerations are recalled in Equation (9.1) (to the aim of conciseness the subscript E for underlining Euler attitude angles is omitted).

$$\begin{bmatrix} \omega_x \\ \omega_y \\ \omega_z \end{bmatrix} = \begin{bmatrix} c_\vartheta \dot{\varphi} - s_\vartheta c_\varphi \dot{\psi} \\ \dot{\vartheta} + s_\varphi \dot{\psi} \\ s_\vartheta \dot{\varphi} + c_\varphi c_\vartheta \dot{\psi} \end{bmatrix}$$

$$\begin{bmatrix} a_x \\ a_y \\ a_z \end{bmatrix} = \begin{bmatrix} -c_\varphi s_\vartheta (\dot{V}_z + g) + c_\vartheta \dot{V}_x + s_\varphi s_\vartheta (\dot{\psi} N_x + \dot{V}_y) \\ s_\varphi (\dot{V}_z + g) + c_\varphi (\dot{\psi} N_x + \dot{V}_y) \\ c_\varphi c_\vartheta (\dot{V}_z + g) + s_\vartheta \dot{V}_x - s_\varphi c_\vartheta (\dot{\psi} N_x + \dot{V}_y) \end{bmatrix} \qquad (9.1)$$

In the considered application, not all the quantities that appear in Equation (9.1) can be measured. In particular, lateral and vertical translational accelerations cannot be measured, while the longitudinal translational acceleration can be obtained applying the derivative operator to the measured longitudinal speed. Thus, the model of the accelerations has to be reduced to

$$\begin{bmatrix} a_x \\ a_y \\ a_z \end{bmatrix} = \begin{bmatrix} -c_\varphi s_\vartheta g + c_\vartheta \dot{V}_x + s_\varphi s_\vartheta \dot{\psi} N_x \\ s_\varphi g + c_\varphi \dot{\psi} N_x \\ c_\varphi c_\vartheta g + s_\vartheta \dot{V}_x - s_\varphi c_\vartheta \dot{\psi} N_x \end{bmatrix}. \qquad (9.2)$$

Now, applying Equations (3.25)-(3.27) to the rotational matrix $R_{ZXY}(\varphi, \vartheta, \psi)$ defined in Equation (7.1), the roll rate $\dot{\varphi}$, the pitch rate $\dot{\vartheta}$ and the yaw rate $\dot{\psi}$ can be expressed as a function of the measured angular rates in the body coordinate system:

$$\begin{bmatrix} \dot{\varphi} \\ \dot{\vartheta} \\ \dot{\psi} \end{bmatrix} = \begin{bmatrix} c_\vartheta & 0 & -s_\vartheta \\ t_\varphi s_\vartheta & 1 & -t_\varphi c_\vartheta \\ -s_\vartheta/c_\varphi & 0 & c_\vartheta/c_\varphi \end{bmatrix} \begin{bmatrix} \omega_x \\ \omega_y \\ \omega_z \end{bmatrix}, \qquad (9.3)$$

and substituting in Equation (9.2), the measured accelerations can be expressed as a function of known signals:

$$
\begin{bmatrix} a_x \\ a_y \\ a_z \end{bmatrix} = \begin{bmatrix} -c_\varphi s_\vartheta g + c_\vartheta \dot{V}_x + s_\varphi s_\vartheta \left(-s_\vartheta/c_\varphi \, \omega_x + c_\vartheta/c_\varphi \, \omega_z \right) V_x \\ s_\varphi g + c_\varphi \left(-s_\vartheta/c_\varphi \, \omega_x + c_\vartheta/c_\varphi \, \omega_z \right) V_x \\ c_\varphi c_\vartheta g + s_\vartheta \dot{V}_x - s_\varphi c_\vartheta \left(-s_\vartheta/c_\varphi \, \omega_x + c_\vartheta/c_\varphi \, \omega_z \right) V_x \end{bmatrix} . \tag{9.4}
$$

The contribution of accelerations that are neglected in Equation (9.4) are recalled:
- *Translational lateral acceleration*: effect of the sideslip of the vehicle, [5], [52], [101];
- *Translational vertical acceleration*: contribution of the heave dynamic of the vehicle and COG elevation;
- *Angular accelerations*: centrifugal and tangential contribution of the accelerations due to roll rate $\dot{\varphi}$ and pitch rate $\dot{\vartheta}$, tangential contribution of yaw rate $\dot{\psi}$;
- *Displacement*: the difference between the position of the COG of the vehicle and the mounting position of the accelerometers is neglected.

To show that Equation (9.2) is a nice description of the measured signals, the expressions in Equation (9.1) and (9.2) are compared using simulated data. The cost function J_A is defined as

$$
J_A = \frac{mse\left(\sqrt{a_x^2 + a_y^2 + a_z^2} - \sqrt{a_{x,M}^2 + a_{y,M}^2 + a_{z,M}^2} \right)}{mse\left(\sqrt{\sqrt{a_x^2 + a_y^2 + a_z^2}} \right)} \tag{9.5}
$$

where the subscript M define the component defined by the considered model. The results are summarized in Table 9.1 and it can be concluded that the introduced approximation is not critical (the data shown in Figure 7.5 have been considered).

	J_A
Equation (9.1)	2.2%
Equation (9.2)	2.3%

Table 9.1: Comparison of accelerations model.

Moreover, considering the model in Equation (9.2), it can be noticed that:
- $J_{A,x} = mse\left(a_x - a_{x,M} \right)/mse\left(a_x \right) = 0.64\%$;
- $J_{A,y} = mse\left(a_y - a_{y,M} \right)/mse\left(a_y \right) = 79.47\%$;
- $J_{A,z} = mse\left(a_z - a_{z,M} \right)/mse\left(a_z \right) = 0.37\%$;

thus, the model commits the greatest error in the description of the nominal acceleration measured along the pitch axis of the motorcycle.

Considering Equations (9.1)-(9.4), the definition of the model to be considered in the Kalman filtering framework is not trivial. As it has been underlined in Chapter 2, many work has been done in the attitude estimation field considering IMU and auxiliary sensors that provide information that are necessary to estimate the attitude of the vehicle and the bias of the sensors [33], [102], [97]. In this work just inertial sensors and speed measurement are available, then, new models have to be proposed.

The basic problem is the definition of inputs, outputs and state variables of the model on which the Kalman Filter is based. In this work the interest is focused on the estimation of the attitude angles, then they will be always considered as state variables. Three possible solutions to define the input and output variables are taken into account and they are shown in Figure 9.1.

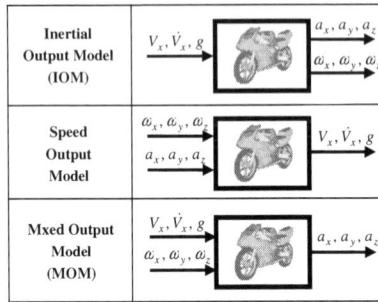

Figure 9.1: Proposed model for Kalman filtering.

9.1.1 Inertial Output Model

In the *Inertial Output Model* (IOM), the acquired inertial signals are considered as outputs of the model while the speed, the longitudinal translational acceleration and the gravitational acceleration are considered as input.

Therefore, the state variables of the model are:

- Euler attitude angle: φ, ϑ, ψ

- Euler angular rate: $\dot{\varphi}$, $\dot{\vartheta}$, $\dot{\psi}$

and, by Equation (9.1) and (9.2), the model of the process can be expressed as

$$\begin{bmatrix} \dot{\varphi} \\ \dot{\vartheta} \\ \dot{\psi} \\ \ddot{\varphi} \\ \ddot{\vartheta} \\ \ddot{\psi} \end{bmatrix} = \begin{bmatrix} 0 & 0 & 0 & 1 & 0 & 0 \\ 0 & 0 & 0 & 0 & 1 & 0 \\ 0 & 0 & 0 & 0 & 0 & 1 \\ 0 & 0 & 0 & 0 & 0 & 0 \\ 0 & 0 & 0 & 0 & 0 & 0 \\ 0 & 0 & 0 & 0 & 0 & 0 \end{bmatrix} \begin{bmatrix} \varphi \\ \vartheta \\ \psi \\ \dot{\varphi} \\ \dot{\vartheta} \\ \dot{\psi} \end{bmatrix} + \begin{bmatrix} 0 \\ 0 \\ 0 \\ \tau_{\varphi} \\ \tau_{\vartheta} \\ \tau_{\psi} \end{bmatrix} + \begin{bmatrix} \eta_{\varphi} \\ \eta_{\vartheta} \\ \eta_{\psi} \\ \eta_{\varphi} \\ \eta_{\vartheta} \\ \eta_{\psi} \end{bmatrix} \qquad (9.6)$$

$$
\begin{bmatrix} a_x \\ a_y \\ a_z \\ \omega_x \\ \omega_y \\ \omega_z \end{bmatrix} = \begin{bmatrix} c_\varphi s_\vartheta g + s_\varphi s_\vartheta \dot\psi V_x + c_\vartheta \dot V_x \\ -s_\varphi g + c_\varphi \dot\psi V_x \\ -c_\varphi c_\vartheta g - s_\varphi c_\vartheta \dot\psi V_x + s_\vartheta \dot V_x \\ c_\vartheta \dot\varphi - s_\vartheta c_\varphi \dot\psi \\ \dot\vartheta + s_\varphi \dot\psi \\ s_\vartheta \dot\varphi + c_\varphi c_\vartheta \dot\psi \end{bmatrix} + \begin{bmatrix} \eta_{a_x} \\ \eta_{a_y} \\ \eta_{a_z} \\ \eta_{\omega_x} \\ \eta_{\omega_y} \\ \eta_{\omega_z} \end{bmatrix}
\tag{9.7}
$$

where the signals η_i are independent zero mean Gaussian white noise process such that

$$
E\left[\begin{bmatrix} \eta_\varphi \\ \eta_\vartheta \\ \eta_\psi \\ \eta_{\dot\varphi} \\ \eta_{\dot\vartheta} \\ \eta_{\dot\psi} \end{bmatrix} \begin{bmatrix} \eta_\varphi \\ \eta_\vartheta \\ \eta_\psi \\ \eta_{\dot\varphi} \\ \eta_{\dot\vartheta} \\ \eta_{\dot\psi} \end{bmatrix}^T \right] = Q \quad E\left[\begin{bmatrix} \eta_{a_x} \\ \eta_{a_y} \\ \eta_{a_z} \\ \eta_{\omega_x} \\ \eta_{\omega_y} \\ \eta_{\omega_z} \end{bmatrix} \begin{bmatrix} \eta_{a_x} \\ \eta_{a_y} \\ \eta_{a_z} \\ \eta_{\omega_x} \\ \eta_{\omega_y} \\ \eta_{\omega_z} \end{bmatrix}^T \right] = R \quad E\left[\begin{bmatrix} \eta_\varphi \\ \eta_\vartheta \\ \eta_\psi \\ \eta_{\dot\varphi} \\ \eta_{\dot\vartheta} \\ \eta_{\dot\psi} \end{bmatrix} \begin{bmatrix} \eta_{a_x} \\ \eta_{a_y} \\ \eta_{a_z} \\ \eta_{\omega_x} \\ \eta_{\omega_y} \\ \eta_{\omega_z} \end{bmatrix}^T \right] = 0
\tag{9.8}
$$

and τ_φ, τ_ϑ and τ_ψ are auxiliary signals that represent the effect of the equilibrium of the moments along the roll, pitch and yaw axis of the vehicle respectively. These signals can be defined as

$$
\begin{bmatrix} \tau_\varphi \\ \tau_\vartheta \\ \tau_\psi \end{bmatrix} = W(s) \begin{bmatrix} \eta_{\tau_\varphi} \\ \eta_{\tau_\vartheta} \\ \eta_{\tau_\psi} \end{bmatrix}
\tag{9.9}
$$

where $W(s)$ is a matrix of weight transfer functions and η_{τ_φ}, η_{τ_ϑ} and η_{τ_ψ} are zero mean Gaussian white noise such that

$$
E\left[\begin{bmatrix} \eta_{\tau_\varphi} \\ \eta_{\tau_\vartheta} \\ \eta_{\tau_\psi} \end{bmatrix} \begin{bmatrix} \eta_{\tau_\varphi} \\ \eta_{\tau_\vartheta} \\ \eta_{\tau_\psi} \end{bmatrix}^T \right] = T.
\tag{9.10}
$$

The IOM description of the process presents many drawbacks:
- The dimension of the state space is six and it is too high for the computational capability of the ECU of a motorcycle;
- Many parameters need to be tuned (*i.e.* Q, R, T and W(s)) and there is not a priori knowledge of feasible values;
- The description depends on the vehicle and on the mounting position, because the auxiliary signals τ_φ, τ_ϑ and τ_ψ would not be general for all the application environment.

An alternative way of description of the process is to consider the inertial signals as inputs to the model and the signals deduced by the longitudinal speed and the gravitational acceleration as outputs.

9.1.2 Speed Output Model

Now consider the *Speed Output Model* (SOM) description of the process in Figure 9.1.

The inertial signals are considered as output, while the quantities related to the longitudinal speed and the gravitational acceleration are the outputs of the system.

Referring to Equation (9.3), the roll rate, pitch rate and yaw rate of the vehicle can be expressed as function of the measured angular rates. This expression can be considered as the state space model of the system and the dimension of the state space is three.

From Equation (9.2)

$$
\begin{bmatrix} a_x \\ a_y \\ a_z \end{bmatrix} = \begin{bmatrix} -c_\varphi s_\vartheta & c_\vartheta & s_\varphi s_\vartheta \left(-s_\vartheta/c_\varphi\, \omega_x + c_\vartheta/c_\varphi\, \omega_z \right) \\ s_\varphi & 0 & c_\varphi \left(-s_\vartheta/c_\varphi\, \omega_x + c_\vartheta/c_\varphi\, \omega_z \right) \\ c_\varphi c_\vartheta & s_\vartheta & -s_\varphi c_\vartheta \left(-s_\vartheta/c_\varphi\, \omega_x + c_\vartheta/c_\varphi\, \omega_z \right) \end{bmatrix} \begin{bmatrix} g \\ \dot V_x \\ V_x \end{bmatrix}
$$
(9.11)

thus

$$
\begin{bmatrix} g \\ \dot V_x \\ V_x \end{bmatrix} = \begin{bmatrix} -c_\varphi s_\vartheta & c_\vartheta & s_\varphi s_\vartheta \left(-s_\vartheta/c_\varphi\, \omega_x + c_\vartheta/c_\varphi\, \omega_z \right) \\ s_\varphi & 0 & c_\varphi \left(-s_\vartheta/c_\varphi\, \omega_x + c_\vartheta/c_\varphi\, \omega_z \right) \\ c_\varphi c_\vartheta & s_\vartheta & -s_\varphi c_\vartheta \left(-s_\vartheta/c_\varphi\, \omega_x + c_\vartheta/c_\varphi\, \omega_z \right) \end{bmatrix}^{-1} \begin{bmatrix} a_x \\ a_y \\ a_z \end{bmatrix}
$$
(9.12)

and finally

$$
\begin{bmatrix} g \\ \dot V_x \\ V_x \end{bmatrix} = \begin{bmatrix} -c_\varphi s_\vartheta & s_\varphi & c_\varphi c_\vartheta \\ c_\vartheta & 0 & s_\vartheta \\ s_\varphi c_\vartheta/\dot\psi & c_\varphi/\dot\psi & -s_\varphi s_\vartheta/\dot\psi \end{bmatrix} \begin{bmatrix} a_x \\ a_y \\ a_z \end{bmatrix}.
$$
(9.13)

The main drawback of Equation (9.13) is that it has a singularity condition when $\dot\psi \cong 0$, that means that when the motorcycle is in a straight condition the attitude cannot be correctly estimated.

Then, the SOM description has the advantage of using a small number of state variables, but it cannot be adopted to estimate the attitude of the vehicle in a straight condition. This problem can be overcome considering the measured accelerations as outputs of the model and the angular rates as inputs.

9.1.3 Mixed Output Model

If the system has a *Mixed Output Model* (MOM) description, then, from Equation (9.3) and Equation (9.11) it can be expressed as

$$
\begin{bmatrix} \dot{\varphi} \\ \dot{\vartheta} \end{bmatrix} = \begin{bmatrix} c_\vartheta & 0 & -s_\vartheta \\ t_\varphi s_\vartheta & 1 & -t_\varphi c_\vartheta \end{bmatrix} \begin{bmatrix} \omega_x \\ \omega_y \\ \omega_z \end{bmatrix} + \begin{bmatrix} \eta_\varphi \\ \eta_\vartheta \end{bmatrix}
$$

$$
\begin{bmatrix} a_x \\ a_y \\ a_z \end{bmatrix} = \begin{bmatrix} -c_\varphi s_\vartheta & c_\vartheta & s_\varphi s_\vartheta\left(-s_\vartheta/c_\varphi \omega_x + c_\vartheta/c_\varphi \omega_z\right) \\ s_\varphi & 0 & c_\varphi\left(-s_\vartheta/c_\varphi \omega_x + c_\vartheta/c_\varphi \omega_z\right) \\ c_\varphi c_\vartheta & s_\vartheta & -s_\varphi c_\vartheta\left(-s_\vartheta/c_\varphi \omega_x + c_\vartheta/c_\varphi \omega_z\right) \end{bmatrix} \begin{bmatrix} g \\ \dot{V}_x \\ V_x \end{bmatrix} + \begin{bmatrix} \eta_{a_x} \\ \eta_{a_y} \\ \eta_{a_z} \end{bmatrix},
$$

(9.14)

where $\begin{bmatrix} \varphi & \vartheta \end{bmatrix}^T$ is the state vector, $\begin{bmatrix} \omega_x & \omega_y & \omega_z & g & \dot{V}_x & V_x \end{bmatrix}^T$ is the input vector, $\begin{bmatrix} a_x & a_y & a_z \end{bmatrix}^T$ is the output vector and the Gaussian white noises with zero mean η_i are such that:

$$
E\left[\begin{bmatrix} \eta_\varphi \\ \eta_\vartheta \end{bmatrix}\begin{bmatrix} \eta_\varphi \\ \eta_\vartheta \end{bmatrix}^T\right] = Q \quad E\left[\begin{bmatrix} \eta_{a_x} \\ \eta_{a_y} \\ \eta_{a_z} \end{bmatrix}\begin{bmatrix} \eta_{a_x} \\ \eta_{a_y} \\ \eta_{a_z} \end{bmatrix}^T\right] = R \quad E\left[\begin{bmatrix} \eta_\varphi \\ \eta_\vartheta \end{bmatrix}\begin{bmatrix} \eta_{a_x} \\ \eta_{a_y} \\ \eta_{a_z} \end{bmatrix}^T\right] = 0.
$$

(9.15)

Some comments can be done:
- The yaw Euler angle ψ is not a state variable, because it cannot be observed in the output signals; this is not a limit for the motorcycle application, in fact on the control point of view this is not a fundamental parameter;
- The MOM description has just two state variables;
- Equation (9.14) does not have singularity conditions in the range of variation of the attitude angles for the considered application.

9.1.3.1 Extension to the MOM description of the process
In the MOM description of the process (as in IOM and SOM) the measured signals are considered to be unbiased and aligned in the body coordinate system. The estimation of the offset of the inertial sensors and the alignment in the body frame are performed as described in Chapter 7. In a Kalman filtering framework, the state space of the process description can be augmented to estimate the mounting attitude and the inertial signals errors. It is now proposed how to extend the MOM description of the system: a model to estimate the mounting attitude and a model to estimate the signals errors are separately presented, but they can be easily be combined in a unique description of the process.

In MOM description, the mounting attitude $(\varphi_o, \vartheta_o, \psi_o)$ can be inserted in the system representation and the three static angles can be estimated. The so obtained model is reported in Equation (9.16). The main drawback of this description of the process is the required computational effort to solve the Kalman filtering problem. The estimation of the mounting attitude in a Kalman filtering framework not only requires to consider an higher order system, but it is also necessary to compute more trigonometric functions. As a consequence the description of the process in Equation (9.14) with a pre-computation of the mounting attitude of the IMU via Equation (7.19) is preferred to the system description in Equation (9.16). In the following Sections the robustness of the Kalman Filter based on MOM description of the process with respect to the mounting attitude compensation will be analyzed.

$$\left\{\begin{array}{l}\begin{bmatrix}\dot{\varphi}\\\dot{\vartheta}\end{bmatrix}=\begin{bmatrix}c_{\vartheta}&0&-s_{\vartheta}\\t_{\varphi}s_{\vartheta}&1&-t_{\varphi}c_{\vartheta}\end{bmatrix}\begin{bmatrix}\omega_x\\\omega_y\\\omega_z\end{bmatrix}+\begin{bmatrix}\eta_{\varphi}\\\eta_{\vartheta}\end{bmatrix}\\[2em]\begin{bmatrix}\dot{\varphi}_o\\\dot{\vartheta}_o\\\dot{\psi}_o\end{bmatrix}=\begin{bmatrix}\eta_{\varphi_o}\\\eta_{\vartheta_o}\\\eta_{\psi_o}\end{bmatrix}\\[3em]\begin{bmatrix}a_{x,o}\\a_{y,o}\\a_{z,o}\end{bmatrix}=R_{ZXY}\left(\varphi_o,\vartheta_o,\psi_o\right)\begin{bmatrix}a_x\\a_y\\a_z\end{bmatrix}+\begin{bmatrix}\eta_{a_x}\\\eta_{a_y}\\\eta_{a_z}\end{bmatrix}\end{array}\right.$$

$$\begin{bmatrix}\omega_x\\\omega_y\\\omega_z\end{bmatrix}=R_{ZXY}^{T}\left(\varphi_o,\vartheta_o,\psi_o\right)\begin{bmatrix}\omega_{x,o}\\\omega_{y,o}\\\omega_{z,o}\end{bmatrix}$$

$$\begin{bmatrix}a_x\\a_y\\a_z\end{bmatrix}=\begin{bmatrix}-c_{\varphi}s_{\vartheta}&c_{\vartheta}&s_{\varphi}s_{\vartheta}\left(-s_{\vartheta}/c_{\varphi}\,\omega_x+c_{\vartheta}/c_{\varphi}\,\omega_z\right)\\s_{\varphi}&0&c_{\varphi}\left(-s_{\vartheta}/c_{\varphi}\,\omega_x+c_{\vartheta}/c_{\varphi}\,\omega_z\right)\\c_{\varphi}c_{\vartheta}&s_{\vartheta}&-s_{\varphi}c_{\vartheta}\left(-s_{\vartheta}/c_{\varphi}\,\omega_x+c_{\vartheta}/c_{\varphi}\,\omega_z\right)\end{bmatrix}\begin{bmatrix}g\\\dot{V}_x\\V_x\end{bmatrix}$$

(9.16)

$$\eta_x=\begin{bmatrix}\eta_{\varphi}&\eta_{\vartheta}&\eta_{\varphi_o}&\eta_{\vartheta_o}&\eta_{\psi_o}\end{bmatrix}^T\quad\eta_y=\begin{bmatrix}\eta_{a_x}&\eta_{a_y}&\eta_{a_z}\end{bmatrix}^T$$

$$E\left[\eta_x\eta_x^T\right]=Q\quad E\left[\eta_y\eta_y^T\right]=R\quad E\left[\eta_x\eta_y^T\right]=0$$

In Equation (9.14), also the signals errors can be considered as new state variables, so that they can estimated in a Kalman filtering framework. Considering the description of the acquired signals in Equations (7.3) and (7.13), the signals errors can be added in Equation (9.14). As a consequence the scheme in Figure 7.1 reduces to the scheme in Figure 9.2 in the offset observation and the attitude estimation are performed by the same block.

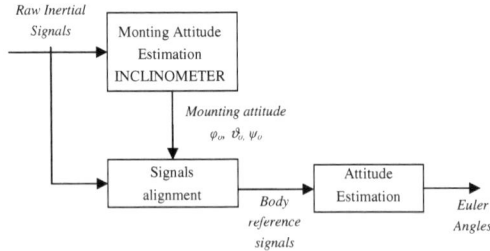

Figure 9.2: High-level estimation scheme without offset observer.

In particular, while the angular rates signals are affected mainly by a bias errors $[\Delta_{\omega_x}\ \Delta_{\omega_y}\ \Delta_{\omega_z}]^T$, the modeled acceleration are characterized by two sources of errors: the bias errors $[\Delta_{a_x}\ \Delta_{a_y}\ \Delta_{a_z}]^T$ and the modeling errors $[E_{a_x}\ E_{a_y}\ E_{a_z}]^T$ due to the approximated model in Equation (9.2). As a consequence, the errors on the accelerations signals are defined by

$$\begin{bmatrix} \varepsilon_{a_x} \\ \varepsilon_{a_y} \\ \varepsilon_{a_z} \end{bmatrix} = \begin{bmatrix} \Delta_{a_x} \\ \Delta_{a_y} \\ \Delta_{a_z} \end{bmatrix} + \begin{bmatrix} E_{a_x} \\ E_{a_y} \\ E_{a_z} \end{bmatrix}. \tag{9.17}$$

The extended MOM representation of the process with inertial signals errors is

$$\begin{bmatrix} \dot{\varphi} \\ \dot{\vartheta} \end{bmatrix} = \begin{bmatrix} c_\vartheta & 0 & -s_\vartheta \\ t_\varphi s_\vartheta & 1 & -t_\varphi c_\vartheta \end{bmatrix} \begin{bmatrix} \tilde{\omega}_x - \Delta_{\omega_x} \\ \tilde{\omega}_y - \Delta_{\omega_y} \\ \tilde{\omega}_z - \Delta_{\omega_z} \end{bmatrix} + \begin{bmatrix} \eta_\varphi \\ \eta_\vartheta \end{bmatrix} \quad \begin{bmatrix} \dot{\varepsilon}_{a_x} \\ \dot{\varepsilon}_{a_y} \\ \dot{\varepsilon}_{a_z} \end{bmatrix} = \begin{bmatrix} \eta_{\dot{\varepsilon}_x} \\ \eta_{\dot{\varepsilon}_y} \\ \eta_{\dot{\varepsilon}_z} \end{bmatrix} \quad \begin{bmatrix} \dot{\Delta}_{\omega_x} \\ \dot{\Delta}_{\omega_y} \\ \dot{\Delta}_{\omega_z} \end{bmatrix} = \begin{bmatrix} \eta_{\Delta_x} \\ \eta_{\Delta_y} \\ \eta_{\Delta_z} \end{bmatrix}$$

$$\begin{bmatrix} a_x \\ a_y \\ a_z \end{bmatrix} = \begin{bmatrix} -c_\varphi s_\vartheta & c_\vartheta & s_\varphi s_\vartheta \left(-s_\vartheta/c_\varphi\left(\tilde{\omega}_x - \Delta_{\omega_x}\right) + c_\vartheta/c_\varphi\left(\tilde{\omega}_z - \Delta_{\omega_z}\right)\right) \\ s_\varphi & 0 & c_\varphi\left(-s_\vartheta/c_\varphi\left(\tilde{\omega}_x - \Delta_{\omega_x}\right) + c_\vartheta/c_\varphi\left(\tilde{\omega}_z - \Delta_{\omega_z}\right)\right) \\ c_\varphi c_\vartheta & s_\vartheta & -s_\varphi c_\vartheta\left(-s_\vartheta/c_\varphi\left(\tilde{\omega}_x - \Delta_{\omega_x}\right) + c_\vartheta/c_\varphi\left(\tilde{\omega}_z - \Delta_{\omega_z}\right)\right) \end{bmatrix} \begin{bmatrix} g \\ \dot{V}_x \\ V_x \end{bmatrix} \tag{9.18}$$

$$\begin{bmatrix} \tilde{a}_x \\ \tilde{a}_y \\ \tilde{a}_z \end{bmatrix} = \begin{bmatrix} a_x \\ a_y \\ a_z \end{bmatrix} + \begin{bmatrix} \varepsilon_{a_x} \\ \varepsilon_{a_y} \\ \varepsilon_{a_z} \end{bmatrix} + \begin{bmatrix} \eta_{a_x} \\ \eta_{a_y} \\ \eta_{a_z} \end{bmatrix}$$

and the stochastic noise variable are such that

$$\eta_x = [\eta_\varphi\ \eta_\vartheta\ \eta_{\dot{\varepsilon}_x}\ \eta_{\dot{\varepsilon}_y}\ \eta_{\dot{\varepsilon}_z}\ \eta_{\Delta_x}\ \eta_{\Delta_y}\ \eta_{\Delta_z}]^T \quad \eta_y = [\eta_{a_x}\ \eta_{a_y}\ \eta_{a_z}]^T$$
$$E[\eta_x \eta_x^T] = Q \quad E[\eta_y \eta_y^T] = R \quad E[\eta_x \eta_y^T] = 0 \tag{9.19}$$

In Section 9.5.1.1, the Extended Kalman Filter based on Equation (9.14) and (9.18) will be compared. Again, the main drawback of the system in Equation (9.18) is that the ECU computational capability is not sufficient for a real time implementation of the estimation algorithm.

9.2 Extended Kalman Filter
In this Section, the EKF for motorcycle attitude estimation is presented. The description of the system based on MOM is used (Equation (9.14)), but it can be easily extended to the extended MOM (Equations (9.16) and (9.18)).
Defining

$$\mathbf{x} = \begin{bmatrix} \varphi & \vartheta \end{bmatrix}^T$$
$$\mathbf{u} = \begin{bmatrix} \omega_x & \omega_y & \omega_z & g & \dot{V}_x & V_x \end{bmatrix}^T,$$
$$\mathbf{y} = \begin{bmatrix} a_x & a_y & a_z \end{bmatrix}^T$$

(9.20)

Equation (9.14) can be written as

$$\dot{\mathbf{x}} = \mathbf{f}(\mathbf{x},\mathbf{u}) + \eta_{\mathbf{x}}$$
$$\mathbf{y} = \mathbf{g}(\mathbf{x},\mathbf{u}) + \eta_{\mathbf{y}}$$

(9.21)

where

$$\mathbf{f}(\mathbf{x},\mathbf{u}) = \begin{bmatrix} c_{x_2} & 0 & -s_{x_2} & 0 & 0 & 0 \\ t_{x_1} s_{x_2} & 1 & -t_{x_1} c_{x_2} & 0 & 0 & 0 \end{bmatrix} \mathbf{u}$$

$$\mathbf{g}(\mathbf{x},\mathbf{u}) = \begin{bmatrix} -c_{x_1} s_{x_2} & c_{x_2} & s_{x_1} s_{x_2} \left(-s_{x_2}/c_{x_1} u_1 + c_{x_2}/c_{x_1} u_3 \right) \\ s_{x_1} & 0 & c_{x_1} \left(-s_{x_2}/c_{x_1} u_1 + c_{x_2}/c_{x_1} u_3 \right) \\ c_{x_1} c_{x_2} & s_{x_2} & -s_{x_1} c_{x_2} \left(-s_{x_2}/c_{x_1} u_1 + c_{x_2}/c_{x_1} u_3 \right) \end{bmatrix} \begin{bmatrix} u_4 \\ u_5 \\ u_6 \end{bmatrix}$$

$$\eta_{\mathbf{x}} = \begin{bmatrix} \eta_\phi & \eta_\vartheta \end{bmatrix}^T \quad \eta_{\mathbf{y}} = \begin{bmatrix} \eta_{a_x} & \eta_{a_y} & \eta_{a_z} \end{bmatrix}^T$$

(9.22)

in which x_i, $i=1,2$, and u_j, $j=1,...,6$, are the *i-th* and *j-th* component of the vectors $\mathbf{x}$ and $\mathbf{u}$ respectively and the noises $\eta_{\mathbf{x}}$ and $\eta_{\mathbf{y}}$ are assumed to be uncorrelated and to have a normal distribution with zero mean and covariance Q and R respectively as defined in Equation (9.15). The filter is implemented in a discrete framework, thus, Equation (9.21) is discretized as

$$\mathbf{x}_k = \mathbf{x}_{k-1} + \mathbf{f}(\mathbf{x}_{k-1},\mathbf{u}_{k-1})Ts + \eta_{x,k}$$
$$\mathbf{y}_k = \mathbf{g}(\mathbf{x}_k,\mathbf{u}_k) + \eta_{y,k}$$

(9.23)

where Ts is the sampling time of the acquired signals and the subscript $k-1$ and k are indicating the sample at *(k-1)-th* and *k-th* instant respectively. In Equation (9.23) the system has been discretized with an Euler forward method. Also some others discretization methods have been considered and it has been verified that the sensitivity to the discretization procedure is low.

The formulation of the EKF in Equation (3.37), can be applied to the model in Equation (9.23). The a priori estimation can be computed with a first order Euler update as

$$\hat{\mathbf{x}}_k(-) = \hat{\mathbf{x}}_{k-1}(+) + \mathbf{f}(\hat{\mathbf{x}}_{k-1}(+),\mathbf{u}_{k-1})Ts$$
$$\hat{\mathbf{y}}_k = \mathbf{g}(\hat{\mathbf{x}}_k(-),\mathbf{u}_k)$$

(9.24)

The linearization of the model evaluated around the most recent estimation can be computed as

$$A_c(\hat{\mathbf{x}},\mathbf{u},k-1) = \frac{\partial \mathbf{f}(\mathbf{x},\mathbf{u})}{\partial \mathbf{x}}\bigg|_{\mathbf{x}=\hat{\mathbf{x}}_{k-1}(+),\mathbf{u}=\mathbf{u}_{k-1}}$$

$$A(\hat{\mathbf{x}},\mathbf{u},k-1) = e^{A_c(\hat{\mathbf{x}},\mathbf{u},k-1)Ts} \cong I + A_c(\hat{\mathbf{x}},\mathbf{u},k-1)Ts + \frac{1}{2}\left(A_c(\hat{\mathbf{x}},\mathbf{u},k-1)Ts\right)^2 \tag{9.25}$$

$$C(\hat{\mathbf{x}},\mathbf{u},k) = \frac{\partial \mathbf{g}(\mathbf{x},\mathbf{u})}{\partial \mathbf{x}}\bigg|_{\mathbf{x}=\hat{\mathbf{x}}_k(-),\mathbf{u}=\mathbf{u}_k}$$

where $A(\hat{\mathbf{x}},\mathbf{u},k-1)$ is the second order truncation of the Taylor series of exponential matrix $e^{A_c(\hat{\mathbf{x}},\mathbf{u},k-1)Ts}$ [103].

Thus, the a posteriori estimation of the attitude of the motorcycle can be computed as

$$P_k(-) = A(\hat{\mathbf{x}},\mathbf{u},k-1)P_{k-1}(+)A(\hat{\mathbf{x}},\mathbf{u},k-1)^T + Q_{k-1}$$
$$K_k = P_k(-)C(\hat{\mathbf{x}},\mathbf{u},k)^T\left[C(\hat{\mathbf{x}},\mathbf{u},k)P_k(-)C(\hat{\mathbf{x}},\mathbf{u},k)^T + R_k\right]^{-1}$$
$$\hat{\mathbf{x}}_k(+) = \hat{\mathbf{x}}_k(-) + K_k\left[\mathbf{y_k} - \hat{\mathbf{y}}_k\right] \tag{9.26}$$
$$P_k(+) = \left[I - K_k C(\hat{\mathbf{x}},\mathbf{u},k)\right]P_k(-)$$

where Q_k and R_k are the tuning parameters of the EKF, $P_k(-)$ and $P_k(+)$ are respectively the a priori and posteriori estimation of the covariance matrix of the state error and K_k is the optimal gain of the Kalman Filter.

The tuning matrices Q_k and R_k will be optimized to obtain the best estimation performance. An alternative way to define Q_k and R_k is by discretization of the matrices Q and R (Equation (9.15)) as

$$Q_k = \int_{\tau=0}^{Ts} e^{A_c\tau}Qe^{A_c\tau}d\tau , \tag{9.27}$$
$$R_k = R$$

then, every a priori knowledge on matrices Q and R can be used to fix Q_k and R_k.

9.3 Unscented Kalman Filter

To the system in Equations (9.21) and (9.22) also the UKF equations reported in Section **3.2.2** can be applied to tackle the motorcycle attitude estimation with inertial signals.

Consider the discrete dynamical system in Equation (9.23), and define the parameters $\alpha=0.4$, $\beta=2$ and $k=2$. Then, define the augmented state vector $\mathbf{x}_k^a = \left[\mathbf{x}_k^T\ \boldsymbol{\eta}_{x,k}^T\ \boldsymbol{\eta}_{y,k}^T\right]^T$ with $L=7$ and the weights $W_i^{(c)}$ and $W_i^{(c)}$, $i=0,...,2L$, defined as in Equation (3.41).

Once the UKF has been initialized as in Equation (3.42), the state variables can be estimated for $k \in [1,...,\infty]$, calculating the sigma points

$$\mathbf{X}_{k-1}^a = \left[\hat{\mathbf{x}}_{k-1}^a \quad \hat{\mathbf{x}}_{k-1}^a + \gamma\sqrt{P_{k-1}^a} \quad \hat{\mathbf{x}}_{k-1}^a - \gamma\sqrt{P_{k-1}^a}\right] \tag{9.28}$$

where $\gamma = \sqrt{L + \lambda}$ and the state vector is estimated as in Equation (9.29) in which the UT transformation presented in Chapter 3 has been applied to the discrete non-linear dynamical system (9.23).

$$\mathbf{X}_{k|k-1}^a = \mathbf{f}\left(\mathbf{X}_{k-1}^x, \mathbf{u}_{k-1}, \mathbf{X}_{k-1}^w\right)$$

$$\hat{\mathbf{x}}_k(-) = \sum_{i=0}^{2L} W_i^{(m)} \mathbf{X}_{i,k|k-1}^x$$

$$P_k(-) = \sum_{i=0}^{2L} W_i^{(c)} \left(\mathbf{X}_{i,k|k-1}^x - \hat{\mathbf{x}}_k(-)\right)\left(\mathbf{X}_{i,k|k-1}^x - \hat{\mathbf{x}}_k(-)\right)^T$$

$$\mathbf{Y}_{k|k-1} = \mathbf{g}\left(\mathbf{X}_{k-1}^x, u_k, \mathbf{X}_{k-1}^v\right)$$

$$\hat{\mathbf{y}}_k = \sum_{i=0}^{2L} W_i^{(m)} \mathbf{Y}_{i,k|k-1} \tag{9.29}$$

$$P_{y_k y_k} = \sum_{i=0}^{2L} W_i^{(c)} \left(\mathbf{Y}_{i,k|k-1} - \hat{\mathbf{y}}_k\right)\left(\mathbf{Y}_{i,k|k-1} - \hat{\mathbf{y}}_k\right)^T$$

$$P_{x_k y_k} = \sum_{i=0}^{2L} W_i^{(c)} \left(\mathbf{X}_{i,k|k-1}^x - \hat{\mathbf{x}}_k(-)\right)\left(\mathbf{Y}_{i,k|k-1} - \hat{\mathbf{y}}_k^-\right)^T$$

$$K_k = P_{x_k y_k} P_{y_k y_k}^{-1}$$

$$\hat{\mathbf{x}}_k(+) = \hat{\mathbf{x}}_k(-) + K_k\left[\mathbf{y}_k - \hat{\mathbf{y}}_k\right]$$

$$P_k(+) = P_k(-) - K_k P_{y_k y_k} K_k^T$$

In what follows, the EKF and UKF are going to be tuned to define the covariance matrices of the noise variables acting on the state and output vectors.

9.4 Filter tuning

In this Section the presented observer are going to be tuned. First of all, the parameters of the EKF are defined, in particular both considering the basic MOM description of the process (Equation (9.14)) and the extension with signals errors estimation (Equation (9.18)). The UKF is just tuned considering the basic MOM description of the process.

9.4.1 EKF tuning

The problem of definition of the covariance matrices of the state equation and output transformation of an EKF is not trivial. In general it is well known that:

- The higher the values of R_k, the smaller the gain of the filter;
- The higher the values of Q_k, the higher the gain of the filter.

Here the covariance matrices are defined using the a priori knowledge of the approximation introduced by the MOM description of the model.

Now consider the system in Equation (9.14). The state equation is based on the kinematic relations imposed by the adopted description of the attitude of the vehicle, thus no approximation is introduced. Errors model are due to residuals errors of the mounting attitude estimation and of offset compensation.

The output transformation introduces many approximations that have been recalled in the beginning of Section 9.1 and there are also residuals errors of the mounting attitude estimation and of offset compensation. As a consequence, the research of the best KF parameters can be constrained considering that $Q_k \ll R_k$.

To tune the parameters of the filter, the mean square error of the performed estimation is evaluated. In particular, the easiest way of definition of the covariance matrices is

$$Q_k = qI_2$$
$$R_k = rI_3$$
(9.30)

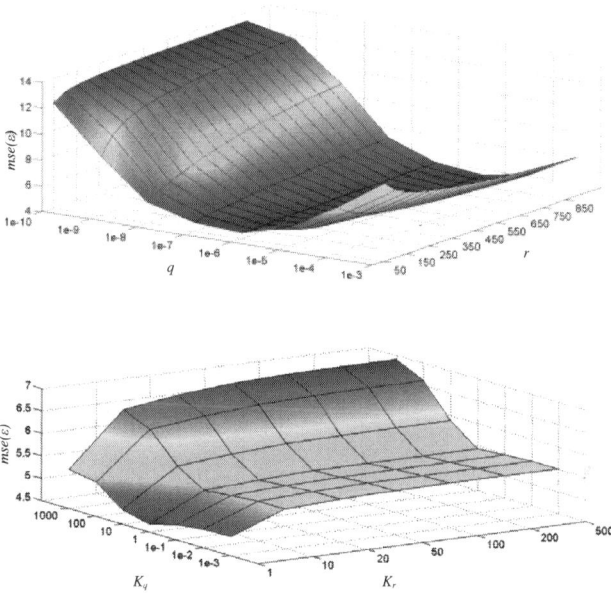

Figure 9.3: Optimization of the EKF parameters for experimental data (minimization of the MSE of the estimation error ε).

Once the parameters q and r are defined so that

$$q = q_o, \quad r = r_o$$
$$q_o, r_o : \min\left\{\sqrt{mse(\varphi - \hat{\varphi})}\right\}$$
(9.31)

it can be further observed that the focus of this application is the estimation of the lean angle of the motorcycle and that the model of the acquired accelerations in Equation (9.2) better fits the signals measured along the x and z axis of the body reference frame, consequently, the variance of the noise signals can be differentiated as:

$$Q_k = q_o \begin{bmatrix} 1 & 0 \\ 0 & K_q \end{bmatrix}$$
$$R_k = r_o \begin{bmatrix} 1 & 0 & 0 \\ 0 & K_r & 0 \\ 0 & 0 & 1 \end{bmatrix}.$$

(9.32)

where the parameters q_o and r_o are fixed and the parameters K_q and K_r are defined so that MSE of the roll angle estimation is minimized.

To define the EKF parameters in the simulation environment, the tracks in Figure 8.6 are covered in general conditions, the estimation error is computed and the mean of the MSE of each simulation is minimized as a function of q, r, K_q and K_r. In the experimental context, all the runs in Misano and Imola are considered, for each of them the estimation error and the corresponding MSE are calculated and the mean of the error variance is minimized.

In Figure 9.3, the mean of the MSE of the estimation error for the experimental test is depicted as a function of the parameters of the EKF; it is shown that:

- The best performance are reached considering $q \in \left[10^{-5}, 10^{-6}\right]$, and in this range the sensitivity to r is very low;
- The best performance are reached considering $K_q = 1$ and $K_r = 1$.

The results of the tuning procedure are reported in Table 9.2.

Parameter	Test Condition	
	Simulation	Experimental test
Q	10^{-3} [rad/s]^2	10^{-6} [rad/s]^2
R	100 [m/s^2]^2	375 [m/s^2]^2
K_q	1	1
K_r	1	1

Table 9.2: Extended Kalman Filter parameters for basic MOM representation of the process.

Now consider the extended description in Equation (9.18). To tune the system the following structure of the covariance matrices has been taken onto account:

$$Q_k = diag\left(q_0, q_0, q_\varepsilon, q_\varepsilon, q_\varepsilon, q_\Delta, q_\Delta, q_\Delta\right)$$
$$R_k = diag\left(r_0, r_0, r_0\right)$$

(9.33)

The parameters q_0, q_Δ, q_ε and r_0 are chosen adopting the following procedure:

0. Fix a value for q_Δ and q_ε;

1. Optimize q_0 and r_0 as in Equation (9.31);

2. Fix q_0 and r_0 to the previous values and optimize q_Δ and q_ε so that $mse(\varphi - \hat{\varphi})$ is minimized;

3. Repeat procedure from 1 until convergence.

In Table 9.3 the parameters obtained using the same data sets of Table 9.2 are reported.

	Test Condition	
Parameter	Simulation	Experimental test
q_0	10^{-4} [rad/s]^2	10^{-8} [rad/s]^2
q_Δ	10^{-4} [rad/s]^2	10^{-11} [rad/s]^2
q_ε	10^{-3} [rad/s]^2	$5*10^{-8}$ [rad/s]^2
r_0	10 [m/s^2]^2	10 [m/s^2]^2

Table 9.3: Extended Kalman Filter parameters for MOM representation of the process with signals errors description.

9.4.2 UKF tuning

The UKF applied to the basic MOM description of the process in Equation (9.14) has been tuned as the EKF.

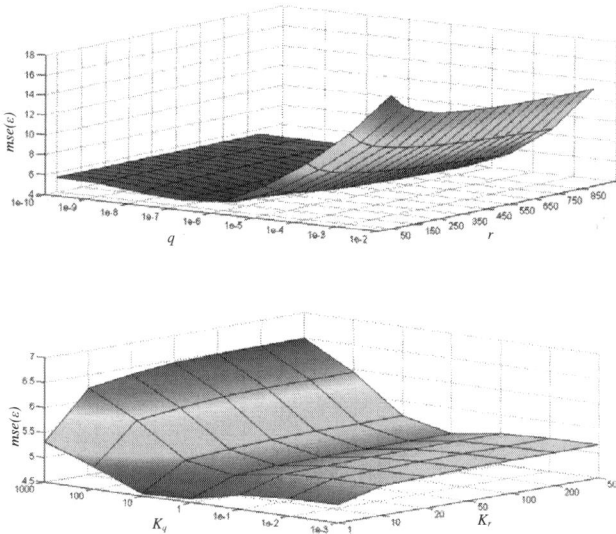

Figure 9.4: Optimization of the UKF parameters for experimental data (minimization of the MSE of the estimation error ε).

First of all the structure of the covariance matrices in Equation (9.30) is considered and $mse(\varepsilon)$, $\varepsilon = \varphi - \hat{\varphi}$, is minimized to obtain q_0 and r_0, thus, the optimization procedure is applied to the matrices in Equation (9.32) to obtain K_q and K_r. In Figure 9.4 the mean of the MSE of the estimation error for the experimental test is depicted as a function of the parameters of the UKF and in Table 9.4 the adopted values for the simulation and experimental framework are summarized.

From Figure 9.4 it can be concluded that for the experimental data:

- $q \in \left[10^{-5}, 10^{-7}\right]$ gives the best performance and in the same time it guarantees the lowest sensitivity to the parameter r;
- K_q and K_r can be set to the unitary value.

In the following Section, the performance of the proposed filters will be compared. A preliminary conclusion can be deduced comparing Figure 9.3 and Figure 9.4: the roll angle estimation accuracy of the EKF and the UKF applied to the basic MOM description of the process are almost the same.

Parameter	Test Condition	
	Simulation	Experimental test
q	10^{-3} [rad/s]^2	10^{-6} [rad/s]^2
r	100 [m/s^2]^2	250 [m/s^2]^2
K_q	1	1
K_r	1	1

Table 9.4: Unscented Kalman Filter parameters for basic MOM representation of the process.

9.5 Performance analysis

Herein, the performances of the Kalman filtering approach for motorcycle attitude estimation are analyzed. The Section is organized as follows:

- In 9.5.1 the EKF and the UKF applied to the basic MOM description of the process are compared, showing that they can achieve the same estimation accuracy; the robustness of the filters to model errors and initial conditions is also taken into account;
- In 9.5.2 the efficacy of the EKF applied to the system description in Equation (9.14) and (9.18) is analyzed and it will be shown that the estimation performance that can be reached with the scheme in Figure 7.1 is comparable to the estimation accuracy of the scheme in Figure 9.2.

9.5.1 EKF and UKF comparison: application to the MOM basic description

Now consider the process described as in Equation (9.14), the parameters of the EKF in Table 9.2 and the parameters of the UKF in Table 9.4: the estimation performance in Table 9.5 are obtained.

Estimation Algorithm	Analyzed Condition	
	General Simulation	Experimental test
EKF	J_{φ_R} 2.1%	J_{φ_R} 2.6%
UKF	J_{φ_R} 2.0%	J_{φ_R} 2.6%

Table 9.5: EKF and UKF estimation performance.

In Table 9.5 the mean performance for simulated and experimental data is reported (J_{φ_R} is defined by Equation (8.40) and the analyzed condition are described in Section 8.3.4) and it can be concluded that the filters have the same estimation accuracy.

9.5.1.1 Simulation results: general condition

In Figure 9.5 the comparison between the reference quantities, the output of the EKF and the output of the UKF is depicted.

It is shown that

- Both the EKF and the UKF do not estimate the lean angle of the motorcycle with respect to the road (φ_R): the observers are able to estimate only the Euler angles that define the rotational matrix $R_{ZXY}(\varphi,\vartheta,\psi)$, thus the bank angle β causes an additive estimation error;

- the EKF and the UKF estimate the Euler pitch angle; the optimization procedure of the filters parameters just consider the estimation performance of φ, the main consequence is that the estimation performance of ϑ_E is not as good as the estimation performance of the roll angle.

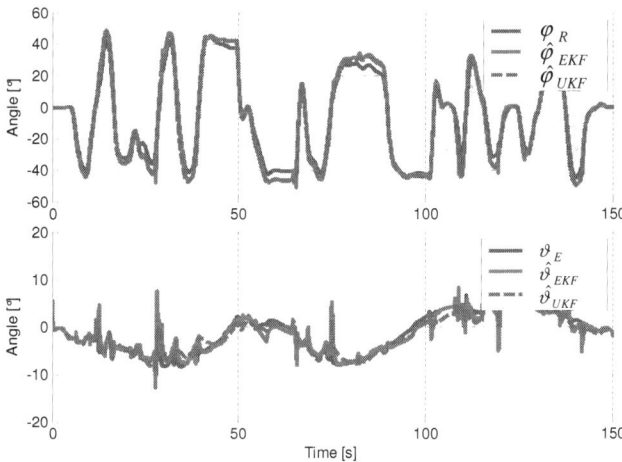

Figure 9.5: Attitude angles estimation with data provided by the simulation of circuit Figure 8.6b in general condition.

It should be not surprising that the performances of the KF are better than the performance of the frequency separation based algorithm. This result can be explained studying the linear correlation coefficient ρ between the estimation error of the optimal observers and the principal quantities that can reduce the estimation performance that are pitch, pitch rate, inertial roll, roll rate and bank.

Estimation algorithm	Linear correlation				
	$|\rho_{\vartheta}|$	$|\rho_{\dot{\vartheta}}|$	$|\rho_{\varphi_i}|$	$|\rho_{\dot{\varphi}}|$	$|\rho_{\beta}|$
EKF	9.3%	2.6%	54.7 %	21.1 %	90.3 %
UKF	7.6%	2.5%	61.1 %	17.9 %	97.9 %

Table 9.6: Linear correlation between estimation error and sources of error for EKF.

From Table 9.6 it can be noticed that the EKF error correlation and the UKF error correlation are comparable. Moreover, comparing Table 8.2 with Table 9.6, it is highlighted that the observer error is less related to the pitch dynamic of the motorcycle then the estimation error of the frequency separation based algorithms. Furthermore, the results in Table 9.6 confirm once more that the road bank is the foremost cause of error in the estimation of the motorcycle lean angle with inertial sensors and that the roll dynamic strongly affects the estimation capability of the observer. The correlation with the roll dynamic is due to the fact that the MOM description of the process is less valid in a dynamical condition than in a static condition.

Finally, it can be concluded that the performance improvement of the EKF end UKF with respect to the frequency separation based algorithms is mainly due to the reduction of the correlation with the pitch dynamic of the vehicle.

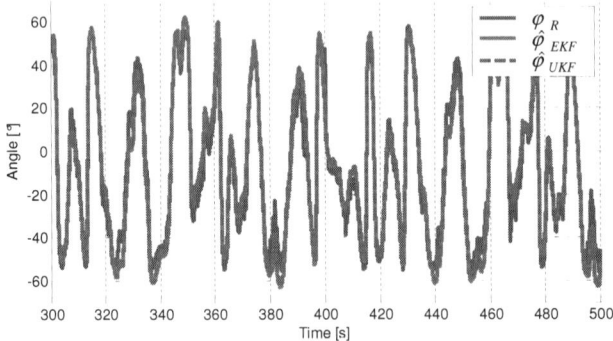

Figure 9.6: Estimation of the roll angle applying the EKF with MOM description of the process with experimental data.

9.5.1.2 Experimental results

Now consider experimental data collected on the Imola circuit, in Figure 9.6 the comparison between the reference roll angle φ_R and the estimated $\hat{\varphi}_{EKF}$ and $\hat{\varphi}_{UKF}$ is depicted. The performance index in Table 9.5 confirms the EKF and UKF efficacy for fitting the roll angle is almost the same, moreover they have a better estimation accuracy than the frequency separation based algorithms.

9.5.1.3 Attitude mounting robustness

Other than the performance improvement, a second great advantage of Kalman filtering is the better robustness to the error of estimation of the mounting attitude of the IMU with respect to the frequency separation based algorithm.

If the static attitude is estimated with an error $(\Delta\varphi_o, \Delta\vartheta_o, \Delta\psi_o)$, then the system should be described by Equation (9.16) with $(\varphi_o, \vartheta_o, \psi_o) = (\Delta\varphi_o, \Delta\vartheta_o, \Delta\psi_o)$. As a consequence the usage of system (9.14) instead of (9.16) introduce a model error.

Two main cases can be distinguished:

- *Pitch mounting inclination error*: now consider that the mounting attitude is $(0, \Delta\vartheta_o, 0)$, thus, from Equations (7.1) and (9.16) the rotational matrix that define the signals measured by the IMU is $R_{ZXY}(\varphi, \Delta\vartheta_o + \vartheta, \psi)$ and the Euler pitch angle estimated by the EKF and by the UKF will be $\hat{\vartheta} = \Delta\vartheta_o + \vartheta$ without reducing the estimation performance of the roll angle (see Figure 9.7 in which the output of the EKF is depicted, similar results can be obtained for the UKF);

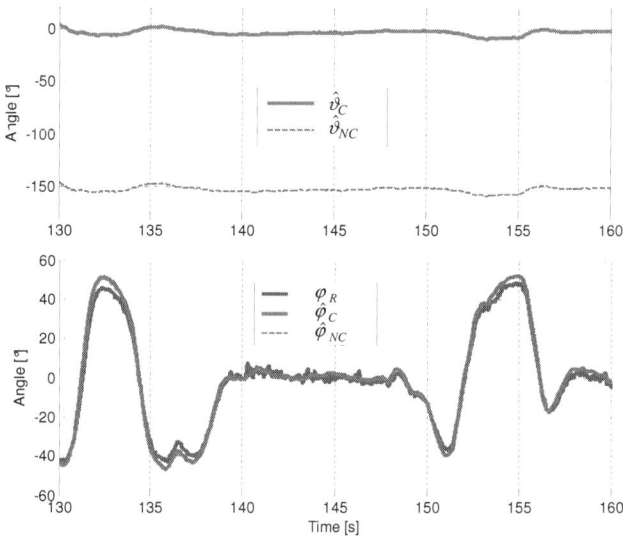

Figure 9.7: EKF output with a mounting attitude $(0°, -148.9°, 0°)$. (a) $\hat{\vartheta}_C$ (pitch angle with compensation) has null mean (dark green line), $\hat{\vartheta}_{NC}$ (pitch angle without compensation) has mean ϑ_0 (dashed red line). (b) The estimated lean angle is not influenced by the mean of the estimated pitch angle.

- *General mounting inclination error*: now consider that the IMU has been aligned to the body coordinate system with an error $(\Delta\varphi_o, \Delta\vartheta_o, 0)$. To analyze the drop off of the performance, experimental data collected on the Imola circuit and Misano circuit are considered and an estimation error of the mounting attitude is simulated. The reduction of the roll angle estimation accuracy is evaluated as ESR/ESR_0 where ESR and ESR_0 are the estimation performance with and without model errors respectively. In Figure 9.8 the mean over all the available data is depicted both for the EKF and the UKF. It can be deduced that the two observers have the same robustness and that the roll angle error $\Delta\varphi_o$ is much more critical then the error $\Delta\vartheta_o$.

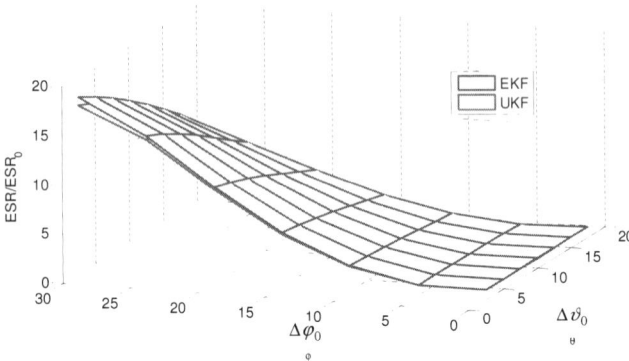

Figure 9.8: Robustness to mounting attitude estimation error of the EKF and the UKF applied to the basic MOM description of the process.

Finally, it can be concluded that, thank to the pitch angle estimation, the proposed approaches based on Kalman filtering are robust with respect to uncertainty of the pitch angle mounting estimation.

9.5.1.4 Robustness to the initial condition

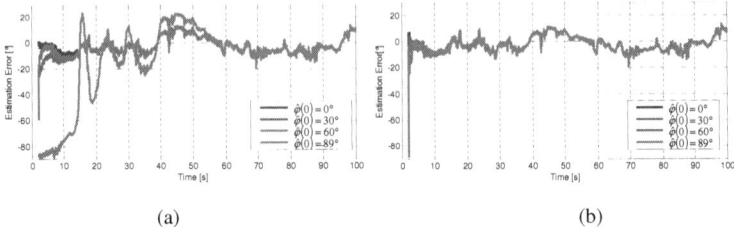

(a)   (b)

Figure 9.9: Robustness to the initial conditions. (a) EKF roll angle estimation error. (b) UKF roll angle estimation error.

It is well known that one of the main advantage of the UKF with respect to the EKF is the robustness to the initial conditions. This conclusion can be deduced also in the application proposed in this Chapter. In Figure 9.9 the estimation output of the EKH and of the UKF are compared for different values of the initial value of $\hat{\varphi}$ and it is highlighted that the UKF has a better robustness to initial condition than the EKF. In the considered application a realistic range of the initial value of the lean angle is ±55° and in this range it is clear that the minor robustness of the EKF implementation is not a critical aspect.

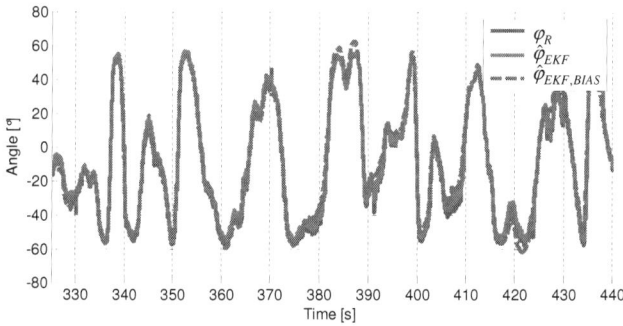

Figure 9.10: Comparison between the estimation output of the scheme in Figure 7.1 and Figure 9.2.

9.5.2 EKF application to the basic and extended MOM description of the process

Herein, the scheme in Figure 7.1 with the EKF applied to the model Equation (9.14) is compared to the scheme in Figure 9.2 with the EKF applied to the model in Equation (9.18).

A lap on the Imola circuit is considered and in Figure 9.10 the comparison between the estimated attitude angles is depicted. It can be concluded that the roll estimation is not influenced by the real-time estimation of the signals errors as state variables of the Kalman Filter; both the ESR of the output estimation of the scheme in Figure 7.1 ($\hat{\varphi}_{EKF}$) and the ESR of the output of the scheme in Figure 9.2 ($\hat{\varphi}_{EKF,BIAS}$) are 2.5%.

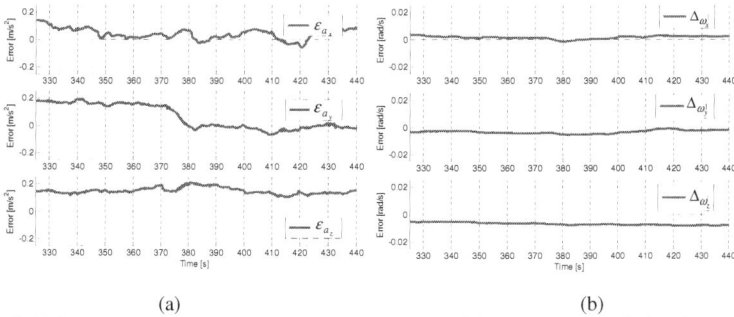

(a)   (b)

Figure 9.11: Errors signals estimated by the scheme in Figure 9.2 by the EKF applied to the system in Equation (9.18).

To the aim of completeness, in Figure 9.11 the estimated signals errors are reported. It can be noticed that the gyroscopes errors signals are almost constant, in fact they are due to the offset of the sensors and they are comparable to the offsets shown in Figure 7.9. On the contrary, the accelerations errors signals are not constants, because they representative also of the models errors introduced by Equation (9.2).

9.6 Kalman filtering and Frequency Split algorithms comparison

In this Section, the performances of the roll angle estimation with the EKF and the Frequency Split (FS) algorithms presented in Chapter 8 are going to be analyzed.

The mains sources of error that are considered are:

- Effect of the tires thickness;
- Effect of the rider's movement;
- Pitch dynamics;
- Road bank angle.

To separate the effect of each of the previous non-ideality, the adopted data are obtained from the BikeSim simulator.

In Figure 9.12, it is shown an application of the FS algorithm to data generated in simplified condition. It is evident the effect introduced by the tires thickness of the vehicle that has been already studied in Chapter 8.

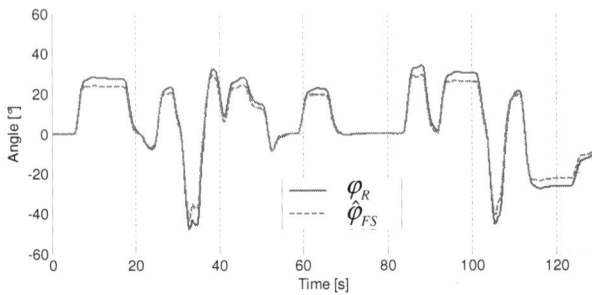

Figure 9.12: Frequency Split algorithm application to data collected in a simplified condition.

As underlined in Chapter 8, the rider's movement tend to compensate the effect of the tires thickness as shown on Figure 9.13.

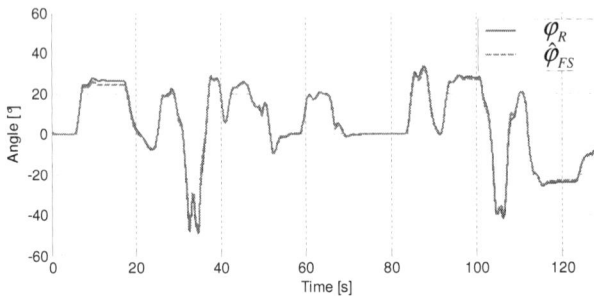

Figure 9.13: Frequency Split algorithm application to data collected with moving rider.

On the contrary the EKF algorithm is not influenced by the tires and the rider: this is the first advantage of the EKF with respect to the FS algorithm.

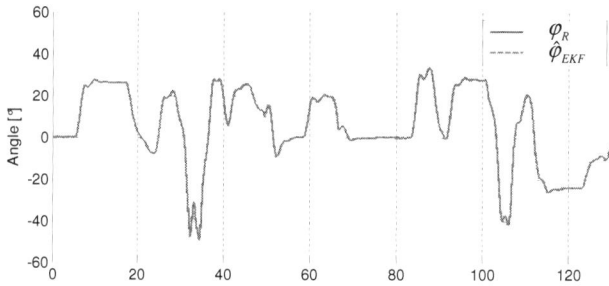

Figure 9.14: Extended Kalman Filter application to data collected with moving rider.

In Figure 9.15, the FS estimator and the EKF are compared using data simulated in a general condition. It can be observed that:

- Between 3 s and 10 s and between 75 s and 85 s a banked road is simulated and both the estimator commit an additive error that is equal to the bank inclination (Chapter 4);
- Between 30 s and 40 s the EKF better follows the reference signal, this is due to the effect of the pitch dynamic of the vehicle (slope of the road and variable speed) that is not compensated in the FS framework.

It can be concluded that the EKF has better performances with respect to the FS algorithms and this is mainly due to the compensation of the pitch dynamics of the vehicle.

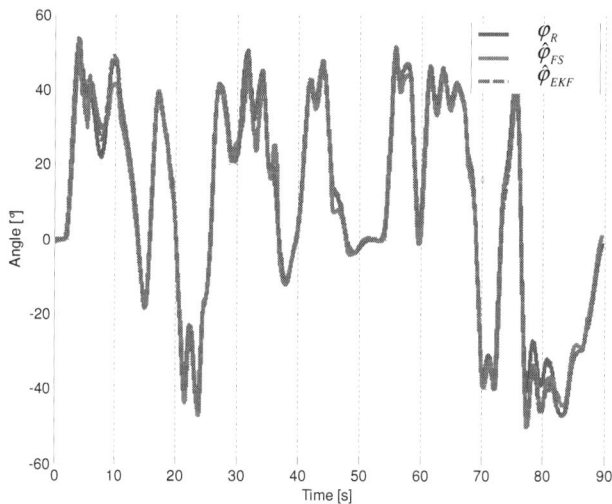

Figure 9.15: Comparison between the FS and EKF estimator with data simulated in general condition.

9.7 Concluding remarks and future works

In this Chapter a Kalman filtering approach for the attitude estimation of a motorcycle has been presented. In particular three solutions have been compared:

1. EKF applied to scheme in Figure 7.1 considering a description of the process in which the only state variables are the attitude angles of the motorbike;

2. UKF based on the model description of Equation (9.14) and pre-compensated inertial signals;

3. EKF applied to the process described in Equation (9.18) in which the signals errors are estimated in real time as state variables of the Kalman Filter.

The MOM description of the process has been introduced: this representation is useful to obtain a simple characterization of the process in which the only state variables are the Euler roll angle and the Euler pitch angle and that is valid in all the possible conditions.

It has been shown that the UKF guarantees a better robustness with respect to the initial condition and that the estimation accuracy is comparable to the performance of the EKF. Moreover, it has been verified that the scheme in Figure 7.1 guarantees the same performance of the approach in Figure 9.2 in which the errors signals are estimated by the attitude observer (EKF).

With respect to the frequency based algorithms, in a Kalman filtering framework it is possible to achieve better estimation accuracy thanks to the pitch angle estimation that guarantees that the identified lean angle is less influenced by the pitch dynamic of the motorcycle.

The EKF based on the MOM description of the process with pre-calibrated inertial signals is currently under test on race vehicles to evaluate the required computational effort on a typical ECU for motorcycle application.

In future works, the aim is the realization of a Kalman Filter to realize the data fusion with inertial sensors and the electro-optical system, so that the attitude angles referred to the road can be estimated with higher precision.

Chapter 10
Conclusions and future works

This book has been devoted to the problem of attitude measurement and estimation in motorcycle applications (Chapter 4-9).

All the proposed estimation problems are far to be closed, but a great step forward has been done. In this Chapter the conclusions and the future works are summarized.

Many modern cars are equipped with a lot of electronic stuffs to control the vehicle dynamic and can be arranged with many different sensors as GPS, IMU, vision systems, wheel encoders, stroke sensors, etc. On the contrary the most recent hyper-sport motorcycles (Aprilia RSV4, Honda CBR, BMW GS1200), are equipped with just ABS system and only the top versions has a traction control system. On the racing field, in the last five years the increasing interest for the vehicle control systems has encouraged the teams to prepare the vehicles with different sensors for telemetry and control purposes.

In particular in motorcycle applications, all the modern control systems such as traction control and stability control, require to be fed by a fundamental quantity: the lean angle of the vehicle.

In Chapter 4 the main definitions of the roll angle are introduced showing which are the differences between the Euler roll angle, the road roll angle and the inertial roll angle. It has been underlined that in a banked road condition the estimation performance quickly degradates and it has been recalled that the inertial roll angle strictly depend on the guidance style of the rider.

The unique system that can be adopted to measure the road attitude angles of the motorcycle can be realized with electro-optical devices. On the market just very expensive and high precision sensors are available to measure the displacement of the chassis with respect to the road. In this book a new low-cost compact sensor has been designed (Chapter 5). It has been shown that, thank to the principle that is used to recover the lean angle of the motorcycle with distance measurements, it is not necessary to employ high precision distance sensors: if the triangulation telemeters are correctly mounted on the motorcycle, a roll angle accuracy less than $1°$ can be reached even if the LASER sensor has an average precision of 1 cm. One of the main problem that has been solved during the system design has been the interaction with the environment: because of the light that is frequency modulated by the asphalt roughness, it is not trivial to obtain an high signal to noise ratio especially in a condition of intense solar light and high speed of the vehicle.

Once the electro-optical system has been designed and deeply tested on racing vehicle both with Moto GP teams and Superbike teams, the problem of generating the reference signal to develop estimation algorithms has been considered to be solved.

In Chapter 6, one of the most frequently adopted system to estimate the motorcycle lean angle has been presented, that is the roll angle estimate via position signals. The principal limits of this method are:

- the lack of the signals in *close sky* condition
- the delay of the GPS system
- the steady state assumption on which the estimation is based that is obviously not true in real applications;
- the estimate is based on the definition of the inertial roll angle, thus, it is the inclination of the motorcycle with respect to the gravity acceleration and not respect to the road.

To improve the performance, the position signals can be combined with inertial signals in an indirect Kalman filter formulation. This results in a very expensive unit that cannot be adopted in racing application nor in the consumer context, and, nevertheless, it cannot recover the attitude parameters of the vehicle with respect to the road.

In Chapter 7-9 the attitude estimation of a motorcycle via inertial sensors for control application has been studied.

First of all, the model of the acquired signals have been discussed to both develop model base algorithm and to support the performance analysis results.

The estimation problem has been addressed solving different tasks separately. Three main topics have been discussed:

- Alignment of the inertial signals in the body frame;
- Signal unbiasing;
- Attitude estimation.

Two main approaches have been proposed and compared to estimate the attitude angles of a motorcycle.

The first one is an easy to implement set of algorithms based on the frequency separation principle (Chapter 8). This algorithms solve the problem of the lean angle estimation of the motorcycle combining high frequency information provided by roll gyroscope, with the low frequency information available from the low frequency components of the acquired accelerations and angular rates. In particular, the described algorithms are based on:

- Vertical and lateral measured angular rates;
- Vertical and lateral acquired accelerations.

The performance of this kind of estimation method has been analyzed in a Neural Network framework that has been useful to give indications for treating the inertial signals and about the main drawbacks of the applied principle. In this analysis and during the experimental test of the algorithms, it has been highlighted that:

- The foremost quantity that as to be derived for estimating the low frequency component of the lean angle is the yaw rate of the motorcycle;
- The hypothesis of null tires thickness degradates the performances;
- The pitch dynamic (inclination of the vehicle added to the slope of the road) strongly affects the estimation accuracy.

It has to be noticed that also the algorithm based on the GPS signals can be integrated in a frequency separation scheme in which the position information are employed to estimate just the low frequency component of the attitude parameter.

The second proposed observer is based on a direct formulation of a Kalman Filter for attitude estimation, in particular both the EKF and the UKF have been designed, discussed and tested (Chapter 9). The process has been described with a Mixed Output Model (MOM) in which the state variables are the roll angle and the pitch angle, the translational longitudinal speed and acceleration together with the acquired angular rates are considered as inputs of the plant and the measured accelerations constitute the observation vector. Also augmented descriptions of the process have been proposed to estimate in the attitude observer also the mounting attitude and the signals errors. The EKF and the UKF have been compared to show that:

- there is not an improvement of the performance due to the unscented transformation;
- the better robustness of the UKF with respect to the initial conditions is not crucial aspect in a motorcycle application.

Implementing the EKF with the augmented MOM description of the process in which also the signals errors are considered in the state vector, it has been shown that the higher complexity of the system is not justified by an higher estimation accuracy.

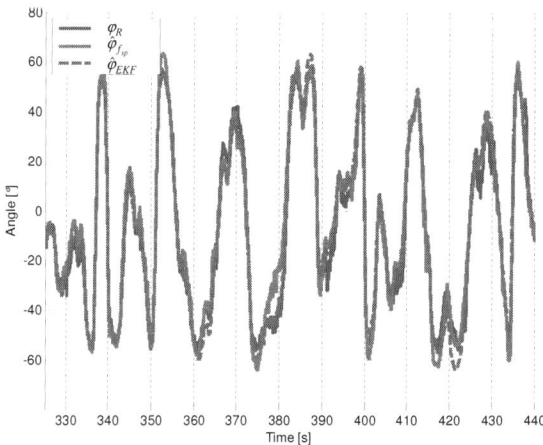

Figure 10.1: Comparison of the lean angle estimation with frequency separation based algorithms ($\hat{\varphi}_{f_{sp}}$, dark green line) and with EKF ($\hat{\varphi}_{EKF}$, dashed red line).

The estimation capability of the proposed attitude observer have been studied both in a simulation environment and with experimental data. The main conclusions are that:

- Banked road is the main cause of error in the roll angle estimation with inertial signals;
- Without estimating the pitch dynamic of the vehicle, as in the frequency based estimation algorithms, the performance of the roll angle estimation reduces;
- The EKF and UKF guarantee, other than a better performance, a greater robustness to uncertainty due to the estimation of the mounting attitude;
- The estimation performance in terms of ESR are around 6% for the frequency separation based algorithms and 2.5% for the Kalman filtering approaches.

In Figure 10.1, the roll angle of the algorithm described in Figure 8.28 is compared with the roll angle estimated by the EKF; a lap on the Imola circuit is considered.The problem of attitude estimation in motorcycle applications is still open. Some new solutions are currently under studying as the description of the inertial data fusion problem as a consensus problem and the combination of information provided by inertial signals and electro-optical instrumentation. Also the application of the proposed algorithms to other racing context such as Power Boat is one of the future developments.

Bibliography

[1] U. Kiencke and L. Nielsen, *Automotive Control Systems for Engine, Driveline, and Vehicle*. Berlin: Springer Verlag, 2000.

[2] S. M. Savaresi, M. Tanelli, and C. Cantoni, "Mixed slip-deceleration control in automotive braking systems," *ASME Transactions: Journal of Dynamic Systems, Measurement and Control*, vol. 129, no. 1, pp. 20-31, 2007.

[3] V. Cossalter, *Motorcycle Dynamics*. Milwakee: Race Dyanmics, 2002.

[4] M. Corno, S. M. Savaresi, M. Tanelli, and L. Fabbri, "On Optimal Motorcycle Braking," *Control Engineering Practice*, vol. 16, no. 6, pp. 644-657, 2008.

[5] H. B. Pacejka, *Tyre and Vehicle Dynamics*. Oxford: Buttherworth Heinemann, 2002.

[6] S. M. Savaresi, M. Tanelli, P. Langthaler, and L. del Re, "New Regressors for the Direct Identification of Tire-Deformation in Road Vehicles via "in-Tire" Accelerometers," *IEEE Transactions on Control System Technology*, vol. 16, no. 4, pp. 769-780, 2008.

[7] R. S. Sharp and D. J. N. Limebeer, "A motorcycle model for stability and control analysis," *Multibody System Dynamics*, vol. 6, p. 123–142, 2001.

[8] L. Gasbarro, A. Beghi, R. Frezza, F. Nori, and C. Spagnol, "Motorcycle trajectory reconstruction by integration of vision and MEMS accelerometers," in *Proceedings of the 43th Conference on Decision and Control*, Nassau, 2004, p. 779–783.

[9] B. Hauser, F.H. Ohm, and G. Roll, "Motorcycle ABS using horizontal and vertical acceleration sensors," US Patent 5,445,44, 1995.

[10] J. K. Schiffmann, "Vehicle roll angle estimation and method," European Patent EP1,346,883, 2003.

[11] P. J. Schubert, "Vehicle rollover sensing using angular accelerometer," US Patent US 1,139,83, 2005.

[12] J. A. Farrel and M. Barth, *The Global Positioning System and Inertila Navigation*. New York, NY: McGraw-Hill, 1999.

[13] E. J. Leffererts, F. L. Markley, and M. D. Shuster, "Kalman Filtering for Spacecraft Attitude

Estimation," *Journal of Guidance, Control and Dynamics*, vol. 5, no. 5, pp. 417-428, 1982.

[14] P. Setoodeth, A. Khayatian, and E. Farjah, "Attitude Estimation By Separate-Bias Kalman Filter-Based Data Fusion," *The Journal of Navigation*, vol. 57, pp. 261-273, 2004.

[15] E. Foxlin, "Inertial head-tracker sensor fusion by a complementary separate-bias Kalman filter," in *Virtual Reality Annual International Symposium*, 1996.

[16] I. Skog and P Handel, "In-car positioning and navigation technologies - a survey," *IEEE Transaction on Inteligent Traspportation Systems*, vol. 10, no. 1, pp. 4-21, March 2009.

[17] M. Norgia, I. Boniolo, M. Tanelli, S. M. Savaresi, and C. Svelto, "Optical Sensors for Real-Time Measurement of Motorcycle Tilt Angle," *IEEE Transactions on Instrumentation & Measurement*, vol. 58, no. 5, pp. 1640-1649, May 2009.

[18] M. Norgia, C. Svelto, I. Boniolo, M. Tanelli, and S.M. Savaresi, "Characterization of Optical Sensors for Real-Time Measurement of Motorcycle Tilt Angles," in *Instrumentation and Measurement Technology Conference Proceedings, 2008. IMTC 2008. IEEE*, 2008, pp. 2070-2073.

[19] I. Boniolo, S. M. Savaresi, and M. Tanelli, "Roll angle estimation in two-wheeled vehicle," *Control Theory & Applications, IET*, vol. 3, no. 1, pp. 20-32, Jan 2009.

[20] I. Boniolo, M. Tanelli, and S. M. Savaresi, "Roll Angle Estimation in Two-Wheeled Vehicles," in *17th IEEE International Conference on Control Applications*, San Antonio, 2008, pp. 31-36.

[21] I. Boniolo, G. Panzani, S. M. Savaresi, A. Scamozzi, and L. Testa, "On the Roll Angle Estimate via Inertila Sensors: Analysis of the Principal Measurament Axes," in *DSCC 2009*, 2009.

[22] M. D. Shuster, "A Survey of Attitude Representations," *The Journal of the Astronautical Sciences*, vol. 41, no. 4, p. 439–517, October–December 1993.

[23] A. A. Shabana, *Dynamics of Multibody Systems*, 2nd ed.: Cambridge University Press, 1998.

[24] H. Cheng and K. C. Gupta, "An Historical Note on Finite Rotations," *ASME Journal of Applied Mathematics*, vol. 56, pp. 139-145, 1989.

[25] L. Euler, *Introductio in analysin infinitorum*. Lausannae, 1748.

[26] R. Pio, "Euler angle transformations," *IEEE Transactions on Automatic Control* , vol. 11, no. 4, pp. 707-715, October 1966.

[27] A. Saccon, "Maneuver Regulation of Non-Linear Systems: The Challenge of Motorcycle Control," Università degli studi di Padova, Padova, Ph.D Thesis 2006.

[28] R. S. Sharp, S. Evangelou, and D. J. N. Limebeer, "Advances in the modelling of motorcycle dynamics," *Multibody System Dynamics*, vol. 12, p. 251–283, 2004.

[29] M. Ang and V. Tourassis, "Singularities of Euler and Roll-Pitch-Yaw representations," *IEEE Transactions on Aerospace and Electronic Systems*, vol. AES-23, no. 3, pp. 317-324, 1987.

[30] S. Haykin, *Kalman Filtering and Neural Networks*.: John Wiley & Sons Inc, 2001.

[31] M. S. Grewal and A. P. Andrews, *Kalman filtering: theory and Practice*.: Prantice-Hall, 1993.

[32] D. K. Arrowsmith and C. M. Place, *Dynamical Systems*. London: Chapman & Hall, 1992.

[33] R. van der Merwe and E. A. Wan, "Sigma-Point Kalman Filters for Integrated Navigation," in *Proceedings of the 60th Annual Meeting of The Institute of Navigation (ION)*, Dayton, OH, 2004.

[34] R. van der Merwe, E. A. Wan, and S. I. Julier, "Sigma-point Kalman filters for nonlinear estimation and sensor-fusion: Applications to Integrated Navigation," in *Proceedings of the AIAA Guidance, Navigation & Control Conference*, 2004.

[35] S. J. Julier and J. K. Uhlmann, "Unscented filtering and nonlinear estimation," *Proceeding of the IEEE*, vol. 92, pp. 401-422, March 2004.

[36] S. Julier and J. Uhlmann, "A New Extension of the Kalman Filter to Nonlinear Systems," in *Proc. SPIE*, 1997, pp. 182-193.

[37] S. Haykin, *Neural Networks and Learning Machine*. Upper Saddle River, NJ: Prentice Hall, 2008.

[38] S. Haykin, *Neural Networks: A Comprehensive Foundation.*: Prentice Hall, 1999.

[39] K. Gurney, *An Introduction to Neural Networks*. London: Routledge, 1997.

[40] R. Rojas, *Neural Network: A Systematic Introduction*. Berlin: Springer, 1998.

[41] S. Bittanti and S. M. Savaresi, "Hierarchically Structured Neural Networks: a way to shape a "magma" of neurons," *International Journal of Franklin Institute*, vol. 335B, no. 5, pp. 929-950, 1998.

[42] K. Hornik, "Approximation Capabilities of Multilayer Feedforward Networks," *Neural Networks*, vol. 4, 1991.

[43] G. P. Zhang, "Neural networks for classification: a survey," *Systems, Man, and Cybernetics, Part C: Applications and Reviews, IEEE Transactions on*, vol. 30, no. 4, pp. 451-462, Nov 2000.

[44] N. Morgan and H. Bourlard, "Generalization and parameter estimation," *Adv. Neural Inform. Process. Syst.*, vol. 2, p. 630–637, 1990.

[45] S. M. Weiss and C. A. Kulilowski, *Computer Systems that Learn*. San Mateo: Morgan Kaufmann, 1991.

[46] A. Weigend, D. Rumelhart, and B. Huberman, "Predicting the future: A connectionist approach," *Int. J. Neural Syst.*, vol. 3, p. 193–209, 1990.

[47] R. Reed, "Pruning algorithms—A survey," *IEEE Trans. Neural Networks*, vol. 4, p. 740–747, Sept 1993.

[48] C. Schittenkopf, F. Deco, and W. Brauer, "Two strategies to avoid overitting in feedforward networks," *Neural Networks*, vol. 10, p. 505–516, 1997.

[49] N. Murata, S. Yoshizawa, and S. Amari, "Network information criterion-determining the number of hidden units for an artificial neural network model," *Neural Networks, IEEE Transactions on*, vol. 5, no. 6, pp. 865-872, Nov 1994.

[50] S. M. Savaresi and C. Spelta, "Mixed Sky-Hook and ADD: Approaching the Filtering Limits

of a Semi-Active Suspension," *ASME Transactions: Journal of Dynamic Systems, Measurement and Control*, vol. 129, no. 4, pp. 382-392, 2007.

[51] S. M. Savaresi, H. Nijmeijer, and G. O. Guardabassi, "On the design of approximate nonlinear parametric controllers," *International Journal of Robust and Nonlinear Control*, vol. 10, pp. 137-155, 2000.

[52] M. Corno, "Active Stability Control Design for Road Vehcles," Politecnico di Milano, Milan, Ph.D Thesis 2008.

[53] M. Vetr, M. Hirsch, and L. del Re, "Curve Safe Traction Control for Racing Motorcycles," *Electronic Engine Controls*, 2009.

[54] M. Tanelli et al., "Control-oriented steering dynamics analysis in sport motorcycles," in *Symposium on System and Identification, to appear*, 2009.

[55] R. S. Sharp, "Stability, control and steering responses of motorcycles," *Vehicle System Dynamics*, vol. 35, p. 291–318, 2001.

[56] R. S. Sharp, "The stability and control of motorcycles," *Journal of Mechanical Engineering Science*, vol. 13, p. 316–329, 1971.

[57] M. Tanelli, M. Corno, I. Boniolo, and S. M. Savaresi, "Active Braking Control of Two-Wheeled Vehicles on Curves," *Submitted to International Journal of Autonoumus Vehicle*.

[58] G. Cocco, *Motorcycle Design and Technology*. Milan: Giorgio Nada Editore, 1999.

[59] Dynasim AB. Dymola-Dynamic modelling laboratory. [Online]. http://www.dynasim.se/dymola.htm

[60] R. Kübler and W. Schiehlen, "Modular simulation in multibody system dynamics," *Multibody System Dynamics*, vol. 4, p. 107–127, 2000.

[61] R. Frezza and Beghi A., "Simulating a motorcycle driver," *New Trends in Nonlinear Dynamics and control, and their Applications*, vol. 295, no. 175-187, 2005.

[62] L. Sass, J. Mcphee, C. Schmitke, P. Fisette, and D. Grenier, "A comparison of different methods for modelling electromechanical multibody systems," *Multibody System Dynamics*, vol. 12, p. 209–250, 2004.

[63] F. Donida, G. Ferretti, S. M. Savaresi, and Tanelli M., "Object-oriented modeling and simulation of a motorcycle," *Mathematical and Computer Modelling of Dynamical Systems*, vol. 14, no. 2, pp. 79-100, 2008.

[64] G. Ferretti, S. M. Savaresi, F. Schiavo, and Tanelli M., "Modelling and simulation of motorcycle dynamics for Active Control Systems Prototyping," in *Proceedings of the 5th MATHMOD Conference*, Vienna, 2006.

[65] M. Tanelli, F. Schiavo, S. M. Savaresi, and Ferretti G., "Object-Oriented Motorcycle Modelling for Control Systems Prototyping," in *Proceedings of the 2006 IEEE International Symposium on Computer-Aided Control Systems Design - CACSD 2006*, Munich, 2006.

[66] S. Donati, *Electro-Optical Instrumentation - Sensing and Measuring with Lasers.*: Prentice Hall, 2004.

[67] M. Norgia, G. Giuliani, and S. Donati, ""Absolute Distance Measurement with Increased Accuracy using Laser Diode Self–Mixing Interferometry in a Closed Loop," *IEEE Trans. on Instrumentation and Measurement*, vol. 56, no. 5, pp. 1894 -1900, 2007.

[68] E. Bava, R Ottoboni, and C Svelto, *Fondamenti della misurazione*, Second edition ed. Bologna, Italia: Società Editrice Esculapio, 2004.

[69] R. Pintelon and J. Schoukens, *System Identification: a Frequency Domain Approach*. New York: IEEE Press, 2001.

[70] ISO, International Standardization Organization, *Guide to the Expression of Uncertainty in Measurement*.: GUM, 1995.

[71] J. Bao and Y. Tsi, *Fundamentals of Global Positioning System Receivers: A Software Approach*. New York, NY: John Wiley & Sons, 2000.

[72] A. El-Rabbany, *Introduction to GPS: the Global Positioning System*. Boston, MA: Artech House, 2002.

[73] E. D. Kaplan and C. J. Hegarty, *Understanding GPS, Principles and Applications*, Second Edition ed. Boston, MA: Artech House, 2006.

[74] B. W. Parkinson and J. J. Jr Spilker, *Global Positioning System: Theory and Application*. Washington: American Instutite of Aeronautics and Astronautics, 1996.

[75] M. S. Grewal, L. R. Weill, and A. P. Andrews, *Global positioning systems, inertial navigation, and integration*.: John Wiley and Sons, 2001.

[76] S. M. Savaresi et al., "Virtual selection of the optimal gear-set in race care," *Int. J. Vehicle Systems Modeling and Testing*, vol. 3, pp. 47-67, 2008.

[77] W. Gander, G. Golub, and R. Strebel, *Least-squares fitting of circles and ellipses*.: Springer, 1994.

[78] V. Cossalter and R. Lot, "A Motorcycle Multy-Body Model for Real Time Simulations Based on The Natural Coordinates Approach," *International Journal of Vehicle Mechanics and Mobility*, vol. 37, pp. 423-447, 2002.

[79] F. Millard, *Principles of Engineering Mechanics*.: Springer, 1986.

[80] H.H. Djambazian, C. Nerguizian, V. Nerguizian, and M. Saad, "3D Inclinometer and MEMS Acceleration Sensors," in *2006 IEEE International Symposium on Industrial Electronics*, 2006, pp. 3338-3342.

[81] D. Lapadatu, S. Habibi, B. Reppen, G. Salomonsen, and T. Kvisteroy, "Dual-axes capacitive inclinometer/low-g accelerometer for automotive applications," in *14th IEEE International Conference on Micro Electro Mechanical Systems*, 2001, pp. 34-37.

[82] Y. K. Peng and M. F. Golnaraghi, "A vector-based gyro-free inertial navigation system by integrating existing accelerometer network in a passenger vehicle," in *Position Location and Navigation Symposium*, 2004, pp. 234-242.

[83] J. H. Chen, S. C. Lee, and D. B. DeBra, "Gyroscope Free Strapdown Inertila Measurment Unit by Six Linear Accelerometers," *Journal of Guidance, control and dynamics*, vol. 17, no.

2, p. 286, March-April 1994.

[84] E. O. Doebelin, *Measurement systems. Application and design.* New York: McGraw-Hill.

[85] G. Xin, Y. Dong, and Z. Gao, "Study on errors compensation of a vehicular MEMS accelerometer," in *IEEE International Conference on Vehicular Electronics and Safety*, 2005, pp. 205-210.

[86] G. Qinglei, L. Huawei, M. Shifu, and H. Jian, "Design of a Plane Inclinometer Based on MEMS Accelerometer," in *International Conference on Information Acquisition, 2007. ICIA '07.*, 2007, pp. 320-323.

[87] Cho and S. Y., "Enhanced Tilt Compensation Method for Biaxial Magnetic Compass," *Electronics Letters*, vol. 41, no. 24, pp. 1323-1325, 2005.

[88] Q. Ladetto, V. Gabaglio, and B. Merminod, "Combining Gyroscopes, Magnetic Compass and GPS for Pedestrian Navigation," in *International Symposium on Kinematic Systems in Geodesy, Geomatics and Navigation (KIS)*, 2001, pp. 205-212.

[89] I. Skog and P. Händel, "A low-cost GPS aided inertial navigation system for vehicle applications," in *Signal Processing XIII: Theories and Applications, Proceedings of EUSIPCO-2005*, Antalaya, 2005.

[90] R. C. Luo, C-C. Yih, and K. L. Su, "Multisensor Fusion and Integration: Approaches, Applications, and Future Research Directions," *IEEE Sensors Journal*, vol. 2, no. 2, pp. 107-119, 2002.

[91] I. Skog and P. Händel, "Calibration of a MEMS inertial measurement unit," in *IMEKO XVIII World Congress*, Rio de Janerio, 2006.

[92] J. C. Hung, J. R. Thacher, and H. V. White, "Calibration of accelerometer triad of an IMU with drifting Z -accelerometer bias," in *Proc. NAECON 1989, IEEE Aerospace and Electronics Conference*, vol. 1, 1989, p. 153 – 158.

[93] A. Kim and M. F. Golnaraghi, "Initial calibration of an inertial measurement unit using an optical position tracking system," in *Proc. PLANS 2004, IEEE Position Location and Navigation Symposium*, 2004, p. 96 – 101.

[94] V. Cossalter, R. Lot, and F. Maggio, "The Influence of Tyres Properties on the Stability of a Motorcycle in Straight Running and in Curve," in *SAE Automotive Dynamics and Stability Conference*, Detroit, Michigan, USA, 2002.

[95] D. H. Titterton and J. L. Weston, *Strapdown inertial navigation technology.*: Institution of Electrical Engineers.

[96] F.L. Lewis, *Applied optimal control and estimation.*: Prentice-Hall, 1992.

[97] J. L. Crassidis and F. Landis Markley, "Unscented filtering for spacecraft attitude estimation," in *AIAA Guidance, Navigation and COntrol Conference and Exhibit*, Austin, 2003, pp. 1-9.

[98] H. Rehbinder and X. Hu, "Drift-free attitude estimation for accelerated rigid bodies," *Automatica*, vol. 40, no. 4, pp. 653-659, 2004.

[99] F. Gustafsson, "Challenges in signal processing for automotive safety systems," in *IEEE*

Statistical Signal Processing Workshop, Bordeaux, 2005.

[100] M. Corno, M. Tanelli, I. Boniolo, and S. M. Savaresi, "Advanced Yaw Control of Four-wheeled Vehicles via Rear Active Differential Braking," in *IEEE Control Systems Society Conference 2009*, 2009, p. submitted.

[101] V. Cossalter, A. Doria, and R. Lot, "Steady Turning of Two-wheeled Vehicles," *Vehicle System Dynamics*, vol. 31, pp. 157-181, 1999.

[102] M. St-Pierre and D. Gingras, "Comparison between the unscented Kalman FIlter and the esxtedend Klaman filter for the position estimation module of an Integrated navigation information system," in *2004 IEEE Inteligent Vehicles Symposium*, Parma, 2004, pp. 831-835.

[103] C. Van Loan, "Computing integrals involving the matrix exponential," *IEEE Transactions on Automatic Control*, vol. 23, no. 3, pp. 395-404, Jun 1978.

11685496R00096

Printed in Great Britain
by Amazon.co.uk, Ltd.,
Marston Gate.